THE ÆSTHETIC MOVEMENT & THE ARTS AND CRAFTS MOVEMENT. EDITED BY PETER STANSKY AND RODNEY SHEWAN. FORTY-EIGHT OF THE MOST IMPORTANT BOOKS, REPRINTED IN THIRTY-EIGHT VOLUMES. GARLAND PUBLISHING, INC.

THE ART OF DECORATIVE DESIGN

CHRISTOPHER DRESSER

GARLAND PUBLISHING, INC.
NEW YORK & LONDON
1977

Bibliographical note:

this facsimile has been made
from a copy in the
Columbia University Libraries
(NK 1510.D817.1862)

Library of Congress Cataloging in Publication Data

Dresser, Christopher.
The art of decorative design.

(The Aesthetic movement & the arts and crafts movement ; 4)
Reprint of the 1862 ed. published by Day and Son, London.
1. Design, Decorative. 2. Design, Decorative--Plant forms. 3. Botany. I. Title. II. Series.
NK1510.D75 1977 745.4 76-17751
ISBN 0-8240-2453-2

PRINTED IN THE UNITED STATES OF AMERICA

THE ART OF

DECORATIVE DESIGN.

THE ART OF

DECORATIVE DESIGN,

WITH

AN APPENDIX, GIVING THE HOURS OF THE DAY AT WHICH FLOWERS OPEN (THE FLORAL CLOCK); THE CHARACTERISTIC FLOWERS OF THE MONTHS (BOTH INDIGENOUS AND CULTIVATED), OF ALL COUNTRIES, AND OF THE DIVERSIFIED SOILS.

BY

C. DRESSER,

PH.D., F.L.S., F.E.B.S.

PROFESSOR OF ORNAMENTAL ART AND BOTANY IN THE CRYSTAL PALACE, SYDENHAM; OF BOTANY APPLIED TO THE FINE ARTS IN THE DEPARTMENT OF SCIENCE AND ART, SOUTH KENSINGTON MUSEUM; AND OF SCIENTIFIC BOTANY IN THE POLYTECHNIC INSTITUTION, AND THE LONDON AND ST. MARY'S MEDICAL COLLEGES.

LONDON:
DAY AND SON, LITHOGRAPHERS TO THE QUEEN,
GATE STREET, LINCOLN'S INN FIELDS.
1862.

THE Author desires to acknowledge the services of the following ladies and gentlemen in the preparation of this work, and to tender them his best thanks:—

MR. W. R. TYMMS,

for the excellent manner in which he has transferred to stone the sketches of the Author;

MR. ARMYTAGE,

of Albion Villas, Barnsbury, N.;

MISS SAWKINS,
MISS EDWARDS,
MISS JULYAN,
MISS STOCK,
MISS McGREGOR,
MISS E. COLLINS,

these ladies being the most talented students of the Schools of Art at Queen-square, Bloomsbury, and the South Kensington Museum, for most valuable assistance in the preparation of the drawings on wood; and

MR. JOHN ALLEN,

a private student of the Author's, for the preparation of the greater portion of the Appendix embracing the Characteristic Plants of the varied lands and soils; hence for this portion of the work this gentleman is more especially responsible.

PREFACE.

THE difficulties under which the Author of these pages laboured when a student have led to the production of the present work. Although studying under men of undoubted talent and kindness, these teachers did not at this time appear to recognise the existence of such general principles in Ornament as could be explained to the student; and this operated so forcibly against the rapid progress of the taught, that upon arriving at a knowledge of the laws which govern the production and combination of ornamental forms through study and research which occupied a period of fifteen years, the Author resolved on setting those general principles before such as desire a knowledge of the Decorative Art which he has observed prevailing in the best ornaments of all ages and has traced in plants, from which he has gained much of his knowledge of true ornament. It is hoped that an acquaintance with the principles set forth in the present work may be as serviceable to the reader as they have been to the Author; if so, that time has not been lost which has been spent in writing these pages.

During the preparation of this work, the writer has been so repeatedly interrupted by manufacturers of the greatest eminence desiring designs for their respective manufactures, that he fears a somewhat disconnected character has been introduced into the work. This has been avoided as far as possible under the circumstances, yet upon this point particularly the Author has to ask indulgence.

North End, Fulham, S.W.
May, 1862.

The student who wishes to pursue his studies in Ornament is strongly recommended to consult Redgrave's "Report on Design," and the "Grammar of Ornament" by Owen Jones, both works of great value.

CONTENTS.

CHAPTER VIII.

CHAPTER IX.

CHAPTER X.

CHAPTER XI.

CHAPTER XII.

CHAPTER XIII.

APPENDIX.

THE

ART OF DECORATIVE DESIGN.

CHAPTER I.

PRIMARILY, ON THE NATURE AND CHARACTER OF ORNAMENT.

1. ORNAMENT is that which, superadded to utility, renders the object more acceptable through bestowing upon it an amount of beauty that it would not otherwise possess: it gives to that which it invests a new charm, as colour bestows upon the flower a new loveliness; and as the colour cannot be said to be essential to the existence of the flower, so the application of ornament to objects cannot be said to be absolutely necessary.

2. Yet this is a principle which is frequently manifest in nature, especially when we call to mind "the merely beautiful appearances of the external world;" for we cannot but think that the functions of the plant could as well be performed, and the works which plants accomplish in relation to the well-being of the kindred creation as fully perfected, were their

shapes coarse instead of rich in refinement, and clumsy instead of delicate and fair; but with no niggard hand has Nature dispensed beneficent gifts, for not only is the mind of man so constituted as to be susceptible of pleasures, but the majority of objects with which we are surrounded are of such a character as administer to delight.

3. Following the dictates of nature and acting in conformity with the spirit of a beautiful creation, we make upon those objects with which we surround ourselves, forms and lines, convolutions and zigzags, and give to them colours and shades which yield pleasure to our minds.

4. In so doing we not only follow the hints given us by nature, but act in accordance with an inward instinct or passion. Man craves after pleasurable emotions, whether he be the untutored savage or a member of the most highly-cultivated society; hence, as certain colours and forms give pleasure to the mind through the agency of the eye, certain sounds through the agency of the ear, and certain flavourings through the agency of the organs of taste; he produces paintings, ornaments, music, and sweets, in obedience to the dictates of his nature.

5. And it is right for him so to do: he is not endowed with a craving nature which is never to be satisfied; with a desire for combinations of forms, colours, and sounds which he is never to produce; but with such desires in order that he may thus be led to produce beauty and enjoy pleasures. The forms which live around him, be they but those of the moss upon the wall and the weed in the sea, are

such as he approves, and are contributions from nature to the delight of her "foster child."

6. Ornament, we have said, is that which beautifies, and renders objects pleasing and delightful; but what is beauty? and whence does it arise? Beauty is that quality of an object which causes delight, gladsomeness, or satisfaction to spring up within the beholder, or induces a thrill of delight in the soul; it arises from the absence of any want, and the presence of that which gratifies.

7. Yet all things which charm the mind are not beautiful. We receive our entire knowledge of external objects through the agency of the nerves; it is these alone which mediate between body and mind, matter and spirit, and inform the inner man of that which is transpiring without. Sounds may charm the ear, fruits have agreeable flavours, flowers most welcome odours, and the downy couch be grateful to the touch; from all of these sources the mind receives pleasure, but the charming the mind through the eye is alone accomplished by the æsthetic or beautiful, hence beauty is that quality of an object which delights the mind through the agency of the eye.

8. From the definition given of ornament it would almost seem that painting, sculpture, and architecture are decorative arts, for they concern themselves with the production of works which are intended to delight the mind through the agency of the eye. In a sense painting, sculpture, and architecture are ornamental arts, and their relationship with each other and with decoration is close, perhaps even more intimate than might be expected until the subject has had much

consideration; indeed, between them there is no marked line of separation, for they meet at their confines as do the kingdoms of nature.

9. A plant which spreads its leaves and unfolds its blossoms in a fixed abode seems very unlike the sportive beast which wanders through the forest at will, yet the animal and the vegetable kingdoms cannot be separated at their lower limits; just so it is with the four arts, sculpture, painting, architecture, and ornament,—they meet at their boundaries, hence their limits cannot be perfectly defined.

10. For the most part ornament is superadded to utility. A wall is a wall, whether decorated or not; and a tube will convey gas equally well whether it has chased upon it beautiful devices or is without enrichment. Not so with painting. As a rule, we value a picture for its intrinsic merit, not as a thing of utility, and not because it has made the canvas more beautiful. We disregard the material on which the scene is produced, and consider it only as that which renders the production of beauty possible; but with ornamentation we value the object decorated because it is now a thing of utility and beauty.

11. We see what ornament is; and the importance of a clear conception and proper understanding at the outset of the nature of decoration or ornamentation cannot be over-esteemed, for unless we have an exact comprehension of the significance of these terms, the time spent in future consideration will be wasted.

12. Our definition of ornament excludes us from a narrow or one-sided view of its character such as we frequently meet with, some contending that the orna-

ment of the Gothic period is alone meritorious, others that with the ancient Greeks true decoration died out, while others maintain that the revival of Classic art in the Renaissance period brought true ornament to light.

13. Few styles of ornament, if any, have arisen which have not possessed some beautiful features. The early Egyptians yet speak to us through their quaint decorations; and in the remains of their departed glory we discover grateful forms and welcome combinations of lines and colours; and the symbolic flower met with that gentle handling and careful consideration at the hands of this ancient people which commands our admiration. In the Classic ornament we read the refinement of the Greeks, and perceive delicacies too subtle for words to express. The Gothic period brought about the fabrication of much beauty, and Renaissance ornament has commendable parts, and neither from the decoration of the Chinese, Indians, or Moors is beauty absent.

14. It is not sufficient that an ornament be pleasing when first viewed, it must give lasting satisfaction. In human society we do not hastily conclude that he is the best man, or even a good man, because he is most pleasing at first; an intimate and prolonged acquaintance may be necessary in order that a right opinion be formed, and *that* man we judge best whose character stands the test of intimacy, and whose company is more pleasurable the longer it is continued.

15. It is so with ornament. That which is at first sight the most pleasing is not necessarily the best, but

that which speaks to us of beauty throughout a long intimacy. Apply this test; ask the question, Does it become more beautiful the longer it is viewed? If it is good ornament it will do so, and will lay hold of the affections, and fasten itself upon them, and its every part will become more and more lovely as it is dwelt upon.

16. Another test of good ornament is this—it powerfully affects the spirits. It can soothe as does sweet music, promote mirth as does the merry air, or hush to reverence as does the solemn anthem. Does it speak joyfully,—it is good ornament. Does it speak solemnly, and yet with well chosen gesture,—it is good ornament. In a thousand strains it speaks, for, like nature, "to the attentive listener it is nowhere dead, never silent."

17. But care must be used in applying these tests. Pleasure is derived from that which is understood; hence, if my knowledge is not sufficient to enable me to comprehend an object, it may seem to me to be without merit, while it is yet most perfect. The nature of the verdict passed upon a work arises, then, from the knowledge of the judge; and, allowing perfect uprightness of character, is right or wrong according to his knowledge.*

18. This being the case, it is obvious that works

* It must ever be remembered that there is no such thing as beauty absolutely. The beautiful is that which satisfies and delights the educated and refined mind, and the most cultivated must be allowed to be the best judges. It is so also in music, for while harsh and discordant notes gratify the untutored savage, subtle harmonies alone give pleasure to the cultivated mind. On this subject the author has set forth his views in *The Builder* of March 15th, 1862.

which to an individual appear beautiful at one time may not at another, and forms which he cannot at first admire he ultimately approves. Indeed, in the history of every art-student a change of this character takes place; the shapes which at first seemed beautiful are ultimately considered coarse, and lines which he could not appreciate he ultimately admires.

19. These considerations reveal to us a folly which is often manifested. Art-critics are numerous; yea, almost as abundant as flies in August. Indeed, every one volunteers an opinion upon an ornament or decoration. I confess my ignorance of music; am I, then, justified in putting forth a critique on an oratorio or opera, and in condemning parts which I do not, through my ignorance, admire? To do so would not be to perpetrate a greater folly than ignorantly to criticise an ornamental form.

20. Applying the tests which we have given for the discovery of the nature of ornament, few objects which are popularly regarded as decorated will be found to be justly entitled to the appellation, and assert their right to the term by manifesting beauty. So utterly incapable of delighting are the majority of our intended wall-decorations, that we do not even seek gratification from them; we use them for covering the nakedness of the wall, but do not for one moment expect that they will produce vibrations on the heart-strings, as do the notes of the lute and the lyre. A few of our old cathedrals awe by their solemn grandeur, and bind in silence the profane tongue as they whisper of their consecration to the service of the Most High;—art is here! The Alhambra,

which is beautifully repeated in the Crystal Palace at Sydenham, overpowers by lavish richness and almost superhuman glory;—art is here! The Greek Court, in the same building, by its exalted refinement; lulls the spirit to sweet reverie—art is here! And when, in St. James's Hall, we appear to be transplanted to some fairer world—art is here! In these last-named places the beauty is excessive, and thereby the spirit is rendered glad, and is entranced by harmonies; the proportions of the parts are just, the enrichments beautiful, and the colours glorious and fair: the whole is a display of knowledge, and learning, and judgment.*

21. It will be found that the amount of pleasure

* The excellency of these works calls loudly to us as a nation to do honour to Mr. Owen Jones, who created them. I would that the nation should use him more while they have him; when it is too late, we shall mourn over our folly. In this we are only like the ancient Greeks, for they rarely valued a man till he was dead; indeed, they first slew him, and then erected a monument to his honour, and I fear that it is not very different with us, for we treat genius coolly, we insult her by a thousand slights, and ultimately starve her, or let her die for want of patronage. It is a strange fact, and yet true, that in this nineteenth century, when philanthropic movements abound, that a Royal Academy student who took the gold medal, and manifested great genius and industry, should die in a garret through starvation, and that many an ornamentist of considerable ability has to retire from the field in order to earn daily bread. Gentlemen, do you like to buy your wives beautiful dresses, when one, two, or at most three shillings was paid for the design which took at least a day to produce, and was the only one chosen out of a score which were offered! And ladies, do you like to sit in your drawing-rooms which are hung with papers for which the designer received but one guinea, and yet the scrolls of beauty and graceful convolutions administer much to your pleasure, remembering that the production of the design took three days, and on the average, only one pattern is sold out of every ten produced.

derivable from the contemplation of an ornament will be largely dependent upon the extent to which mind is embodied in it. Stephenson invested coal with a new interest when he told us that it was buried sun's rays—a lump of heat and light! A strange statement this, yet true; and it is also strange and true that the man, while yet dead, speaks through his works. Man may be said to be a spirit (the body is animal), for it is the mind which exalts him above all creatures amidst which he abides, and constitutes him a superior being. In *this*, man recognises the stamp of exalted humanity, and by it he discovers a brother, whatever be his habits or colour.

22. How delightful it is for mind to converse with mind, and tell of joys experienced and pleasures reaped in order that both may be glad! Sympathy is but mind sharing lot with mind, and love but a bond which unites spirits. This source of exalted gratification can be embodied in ornament; and our pleasure in a form or composition will be, to a great extent, in ratio to the amount of mind which it conceals.

23. We have a manifestation of the pleasure derivable from an embodiment of mind in pictorial art in the works of Martin. Here we have such scenes as mind alone can conceive;—a prison-house more terrible than words can express;—a palace with immeasureable domes, and burning lights poised in the air. We glory in these works, but not because of the perfection of the drawing, for this is often defective, nor through any extraordinary handling of the colours, for in this respect also there is nothing calculated to

command our interest, but because of the mind which they embody.

24. As a rule, the greater the manifestation of mind in a work of art of any description, the more pleasure we derive from it; also, the absence of the manifestation of mechanical labour in its construction tends to the same result. "Those arts," says Sir Charles L. Eastlake, "are generally considered the most worth in which the mental labour employed and the mental pleasure produced are greatest, and in which the manual labour, or labour of whatever kind, is least apparent."

25. We have said that few, if any, styles of ornament have arisen which have not possessed some meritorious or beautiful features. From this it might be thought that it is well to copy the ancient forms and apply them to the purposes of modern decoration, for beauty cannot become old; but this is by no means desirable. Although beauty cannot wear out, a repetition of ancient forms is not appropriate, for ornament, like architecture, must express the sentiments of the age in which it is created.

26. How full of meaning are some of the ancient ornamental forms! The Lotus (fig. 1) not only betokened to the Egyptian coming plenty, but was an expression of their faith, it being to them as a Deity. After the inundation of the Nile, the rice was set by being cast upon the waters, through which it sank into the soft and fertile alluvial deposit. In the waters the Nymphæa* pushed forth its fresh foliage,

* Nymphæa is the botanical generic name for the Egyptian water-lily, or lotus.

which it gradually raised to the surface; and ultimately from amidst the leaves its fragrant blue flowers arose. The appearance of this plant was speedily followed by the springing up of the rice and the succeeding harvest, but the token of the harvest was found in the lotus-lily:—it was this which spoke of coming plenty; of the season of fruitfulness, and of approaching good.

Fig. 1.

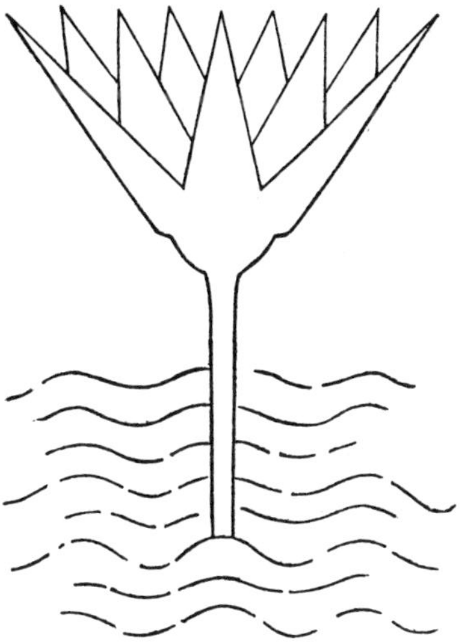

27. Under such circumstances we do not wonder that the appearance of this plant was anxiously looked for; that its development was watched with interest, and that the Egyptian joyed in its flowers. But delight was gradually replaced by reverence; the flower which at first gave pleasure ultimately received the homage and adoration of a nation, and became the fancied abode of a God; and in ratio to the worship paid to the spirit of the flower, and ultimately to the

flower itself, was the abundance of the harvest, according to the faith of the Egyptian.

28. The pleasure taken in the plant by this ancient people is set forth by the frequency of its occurrence in their decorations; and as the cross has been used in later times to prompt to worship, so this flower, even when transmuted into a conventional form, reminded them of rites and prayers. Fully, then, was the lotus-form expressive of the faith and sentiments of the age in which it was created.

29. This is not an exceptional case, but a sample of what true art has ever done. In the Greek ornament there are but few simple forms, and even these do not speak as does the Egyptian lotus. Yet the Greek ornament, as a whole, will be found to give utterance to sentiments in a manner most impressive.

30. Refined forms are an expression of refined minds, and delicate subtleties in shapes indicate a keen power of perception. No system of decoration with which we are acquainted embodies refinement so fully as does the Greek, for it is *rich* in refinement, and is stored with delicacies. But it speaks of more than the refinement of the Greeks; it reveals to us their exalted perception of the beautiful, and this it does so clearly as to render such a manifestation apparent even to the more careless observer. Besides this, every feature, both of an architectural and decorative character, in a Greek edifice, is a revelation of a knowledge of Nature and intimacy with her. To illustrate this by an

architectural member we may select the Doric column (Plate I.)—an example of the most exalted Greek creation, which appears to rise with the energy of vigorous life, and yet seems pressed on by a weight which it has to sustain; the superincumbent mass resting upon the shaft gives to it turgidity, and the capital manifests the same distension. But the acknowledgment of pressure is by no means offensive, for the column evinces strength sufficient for the support of the superposed weight, yea, strength which is evidently more than equal to the work performed; hence the apparent distension of the pillar by weight only results in the formation of curves of extreme delicacy and shapes of subtle beauty. It is so with the Greek ornaments: there is a recognition of the working of natural law in each, and an embodiment of energetic life, and this strife between pressure and life brings about beauty almost inimitable.

31. The decorations of the Greeks express to us their keen knowledge of the laws and perception of the beauties of Nature. In like manner the mediæval ornament has its tale to tell.

32. In the earlier periods of Christian art great purity is manifested, and an amount of simplicity which is very grateful; but even in later times, when the simplicity is to a great extent lost, the ornament speaks the sentiments of the age in which it was created. The trefoil, or three-lobed leaf, speaks of the Trinity; the quatrefoil, or four-lobed leaf, of the four Evangelists; while the passion-flower tells the

whole story of the cross. A thousand other forms also speak through their symbolic character as well as the knowledge which they embody.

33. What could be more impressive as a symbol of the passion of our Lord than the pelican feeding its young with its own blood: here we have a creature picking its breast till blood issues forth, in order to the nourishment of those which proceed from it. Thus we have set forth the blood-shedding of Christ, in order that those whom he created may live.

34. Ornament, then, ever has expressed the sentiments of the age in which it has originated, and the ancient decorative forms still tell of the faith of our ancestors; but as our creed differs from theirs, and as we are not prepared to endorse all their sentiments, we cannot fitly appropriate to ourselves those ornaments with which we do not sympathize, as they are an expression of sentiments in which we cannot concur.

35. It would seem difficult, at this late period of the world's history, to originate a new scheme of ornamentation which should be an expression of sentiments; and should such be possible we might not readily be able to determine what should be expressed. Want of concord retards any expression of religious faith, for unless there is unanimity the views of the nation cannot be set forth by any one system of ornamentation, any more than by one style of architecture.

36. But the Greeks strongly impressed their decoration with a marked character which stamps it at

once "a style," and distinguishes it amidst the ornaments of all other periods, while their religious views were, if possible, more varied and antagonistic than our own.

37. Though the expression of unanimity of religious principle is at present impossible, yet there are sentiments in which all could acquiesce. We seek refinement, as did the Greeks, and we have knowledge which is waiting to be embodied in form; yet neither the Greek ornament nor architecture are appropriate with us, for our knowledge is not as the knowledge of the Greeks, and unless expression, as uttered in decorative forms, is so modified as to accord with the material and climate in which such expression takes place, it fails to fulfil that which is required of it.

38. The putting forth a volume of ancient literature as a new thing would be regarded as a folly, and the right of an individual to present such under this character would be questioned; and, if a few sentences or paragraphs were altered, this would notgreatly lessen the wrong. Yet so it is with both ornament and architecture, which are so nearly associated that in these remarks they cannot be separated; we reproduce heathen temples without number, modifying them to an extent sufficient to render the performance of worship in them possible, or to fit them for the purpose of their erection, and no more; or we repeat an old Christian edifice no better suited to the requirements of our day:—"The nave and aisles of a Gothic church become absurd when filled with pews for Protestant worship, where all are

required to see and hear. The columns of the nave which impede sight and sound, the aisle for processions which no longer exist, rood screens and deep chancels for the concealment of mysteries now no longer such, are all so many useless reproductions which must be thrown aside," and the decorations are no more appropriate.

39. We cannot but think that one great token of a new style of ornament has been afforded us in St. James's Hall, and, to speak figuratively, had we, as a nation, the wisdom to work the mine from which this came, we should not be long before we were sheltered by such buildings as should by their excellency delight us, and, by embodying in form sentiments in which we acquiesce, command our sympathies.

40. That St. James's Hall is beautiful admits of no dispute, but should the style of the decoration be asked we could not reply. It has features which are Egyptian, yet it is not Egyptian; it has parts which are somewhat Greek, yet it is not Greek; and few styles have appeared which are not suggested by some ornamental character, and yet the decoration is neither in any one style, nor in all jumbled together. He who is conversant with all literature manifests his knowledge in his writings, yet he does not reproduce the ancient Classics. So, in St. James's Hall, Mr. Owen Jones has displayed a deep acquaintance with all styles of ornament which have passed away, and has expressed in a new form the refinement of our age, which finds utterance also in sound in this very place.

41. It will be thought that we are dealing out

praise lavishly to this work by Mr. Owen Jones, and that it is inconsistent so to do, as our great monuments of Gothic architecture are worthy of especial commendation. We desire to do justly, and act independently of the smiles of men. The great merits of our truly grand monuments of mediæval art have repeatedly been told, and their excellency is fully acknowledged, hence it would be folly to direct attention to that which has repeatedly been heard; the merits of these great works we most highly appreciate, and have selected St. James's Hall as an illustration, primarily because of its novelty.

42. Many appear to regard ornament as the realization or embodiment of a dream of beauty, and the ornamentist as one who is favoured with a kind of revelation, for, say they, we cannot control our thoughts or compel our minds to originate new ideas. Such a view is founded in error, for we can by cultivation and careful encouragement induce the mind to produce the original, and ornament does not result from an unintelligent dream.

43. Knowledge is the great source of ornament,—it is the rich soil from which spring creations that attract by their beauty and delight by their gracefulness, and the merit of the ornament will be dependent upon the knowledge of its originator.

44. Ornament is obedient to rule and law, and the principles which govern it can be learned and comprehended. The purest ornaments are founded upon some common plan, or are subject to the same common rule, while they appear in many and varied forms. Yet all the laws which govern the production

of ornament are not understood, nor does every member of every successful design occupy its position in obedience to a law known to the producer, for judgment has to decide in many difficult cases; nevertheless it is manifest that knowledge is essential to the creation of the beautiful, as we never find good ornament produced in the absence of great knowledge.

45. The plant grows in obedience to law, and upon fixed principles; it has a determined form and a particular habit of growth; the flower proceeds in all cases from the same part of the plant, and works the production of the same kind of fruit; the seed also germinates under fixed conditions, and after sprouting in a manner pre-determined by law, brings about the repetition of the parent plant. There is no accident here; the flower has not one shape one day, and, as the result of casualty, a new form when next viewed; and by the production of seed the plant produces its like, and not accidentally a rose or a thistle.

46. Yet in some strange and exceptional cases the seed does not bring about the repetition of the parent, and rule seems to be departed from. The seed of the cabbage produces a cauliflower, and of the *Celosia cristata* the cockscomb; but even here an unalterable law is working the change, for it is determined that more than a fixed quantity of mineral matter cannot be taken into the plant-system without a change of form resulting; and even the size of the plant is decided by the amount of heat and light which it receives, together with material food; for as the erection of a building is dependent upon the presence of

material, the expenditure of force in raising the material, and the activity of a ruling power which determines the positions of the parts, so is the plant dependent for its existence upon the presence of befitting matter, the application of power to the raising the matter, and the working of an overruling influence.

47. That the production and development of the plant is governed by rule is apparent, yet we do not trace the law in all cases. It seems arbitrary that one plant should grow in almost any soil and under any sky, as the little daisy, while another is confined to a small island or particular mountain range, as the Wulfenia to the Carinthian Alps. Although we have not yet discovered the entirety of the law on which the distribution of plants depends, we do not doubt its existence, for the secrets of its workings are ever being disclosed to the attentive inquirer; so in ornament we know that much is dependent upon law, and are daily making acquaintance with principles which to us are new, hence we cannot doubt that all beautiful forms are subject to rule. It will be our endeavour to set forth the laws which govern the production of decorative compositions, and to point, as far as we may be able, in the direction that new laws may be sought.

48. We have inquired into the nature of the embellishing art, but what is its value? That which administers to mental gratification cannot be worthless and is not altogether independent of bodily health. Setting aside the ministrations of decorative

art to happiness and the enhancement of joys, and its promptings to social intercourse, it has even a commercial or money value in the busy world.

49. An illustration used by Mr. Wornum, the Keeper of the National Gallery Collections, is here appropriate. He called attention to a plain white jar in which was sold one pound of the best Dundee marmalade for seven-pence. He then exhibited a second utensil, which had a fanciful shape and was impressed with the significant figure of the thistle,—in this was sold one pound of the same marmalade for nine-pence; and lastly, he exhibited a jar which was not only moulded into a particular form, but had coloured oranges upon its sides,—in this one pound of the same exquisite preparation was sold for one shilling. Seven-pence, nine-pence, and a shilling the same quantity of the same kind of marmalade fetched in view of the character of the jar, and yet that which was coloured would not cost one penny more than that which was plain, probably not a halfpenny; hence, through the agency of ornamentation, fourpence more profit, at least, was made upon each pound of marmalade sold.

50. In relation to the example just given it should be noticed that the so-called decoration of either jar was a questionable adornment; the shapes were not pleasing, the forms wrought upon them were not disposed with discretion, neither was the colouring good. If this be the value of bad art, what must the commercial importance of good ornament be?

51. As a principle, it will be found that the value of the manufactured article is dependent upon the

knowledge displayed in the using and adorning the material, and not upon the amount of labour expended upon its construction. The same clay can be wrought into a thing of beauty or an object without comeliness, and the most welcome ornaments are usually both simple in character and sparingly used, for extravagance in ornament is as offensive as extravagance in dress.

CHAPTER II.

THE MINISTRATIONS OF PLANTS TO ORNAMENT.

1. ONE great work of the ornamentist is that of refining his mind in order that he may be enabled to discover and fully appreciate those delicacies of form and line which are unperceived by the untutored and careless observer: to this end the formation of an intimacy with nature will be found most conducive, as it inevitably leads to the cultivation of taste; and the ministrations of plants to the production of the more exalted forms of ornament are in this particular manifest and salutary.

2. In order to this refinement, the wood, the garden, and the hedgerow will each form a suitable studio; for the world of life, especially of plant life, must be constantly observed. The form of the leaf must be investigated; nor is it sufficient that each leaf receive a cursory glance or hasty observation, it must become the subject of much thought and prolonged meditation. A simple leaf will form a good theme for attentive consideration during many hours, and the more anxious inquirer may profitably spend even a week or fortnight in contemplating an unpretending form. Indeed, it is not the number of refined shapes viewed which tends so much to the cultivation of taste as the full understanding of the few. He who

reveals the cause of the beauty of a solitary leaf thereby manifests that he has seen more than he who has hastily viewed the riches of many lands.

3. If a week is not too long to spend in the consideration of a simple leaf, how long should we meditate upon a flower? Say a month. Happy should we be in devoting every hour of this time to the consideration of a solitary flower, if we thought that we had sagacity sufficient to enable us in that time to perceive and fully understand its beauty. We have sat in our vinery for hours gazing at the leaves above as their forms have been sharply defined upon the blue heavens. We have gone evening after evening and day after day to endeavour to discover the source of the fulness of beauty manifested in the form of a solitary leaf, and have failed to comprehend the origin of half its comeliness; if hours, yea weeks, are insufficient for the full understanding of the beauty of a leaf, how much more shall they fail to reveal the matchless loveliness of a flower!

4. The flower! There are many leaves here, and parts varied in magnitude; there are small parts within and large parts without, and members of diversified shapes; there are lines uniting with lines, shapes contrasting with shapes, and colours blending with colours, all of which are worthy of the most attentive study, and he who would refine his taste and cultivate his perception of beautiful form must be content to study a flower till his eyes grow weary with viewing it, and his heart learns to *feel* the beauty which his eyes perceive.

5. Plants minister to ornament, we see, by contri-

buting to the refinement of taste; they also furnish decorative shapes.

6. The vast stores of nature are open to us, and from the floral kingdom, as from a mine of inexhaustible wealth, the ornamentist derives forms of inexpressible beauty. There are the forms of leaves, of flowers, and of fruits, many of which are of rare excellence as ornamental shapes; and besides the innumerable modifications of form seen in the parts named, there are peculiarities of contour in organs of lesser interest. The stem is not always round, but has manifold shapes; the leaf-bud, especially during its unfolding in spring, the flower-bud, and not unfrequently the seed, furnish forms which cannot be too highly appreciated or too fully considered; and here we do not mention the ten thousand beauties revealed by sections and dissections of flowers and plants.

7. To take the herbaceous or succulent (not woody) stem as an illustration of the variety presented by a member which is commonly regarded as without interest ornamentally, we find that it is triangular in the sand-grass (*Carex*, fig. 2, *e*), and papyrus (the bulrush of the Scriptures, fig. *f*); in the former the sides are concave and the angles acute, in the latter the sides are convex and the angles are rounded. In other instances the stem is square (*i*), but upon the quadrangular form a number of modifications are wrought, like changes of note upon the lyre with one string; in some instances the angles are replaced by flat faces (*j*); in others by rounded surfaces, or little circular ridges (*n*). In the fiddle-plant of the country children, the *Scrophularia nodosa* (*l*), the sides are

slightly hollowed, the angles rounded, and each corner is furnished with a membranous ridge; in the common stinging-nettle the square is modified into a very beautiful form (*o*). We have hexagonal stems

FIG. 2.

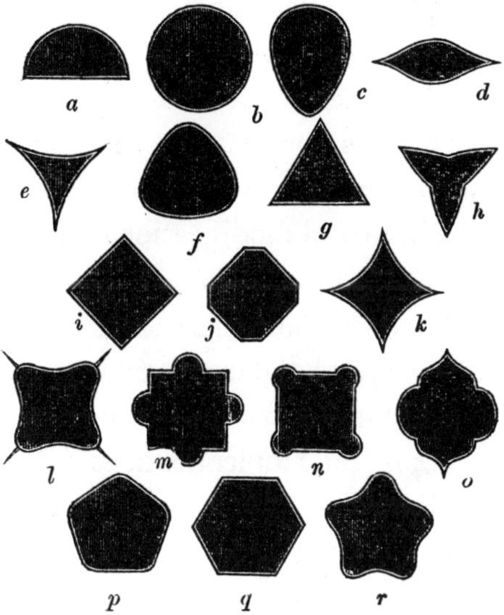

in the hop and garden balsam (*q*); besides which there are egg-shaped (*c*), biconvex (*d*), semicircular (*a*), and pentagonal forms (*p*, *r*).

8. Actual forms, we say, are furnished by nature; and, we may add, forms of exceeding beauty. Who can conceive a form fairer to look upon than the lily, or a shape more gentle than the violet? The lowly grass, by its gracefulness, bids defiance to imitation, even by the keenest observer and most careful portrayer; and the weed in the sea has charms which are unfolded to those alone who attentively observe it for years.

9. Plants not only furnish the ornamentist with

abstract forms, but are materials brought forth for the use of the beautifier; we mean by this, that as stone is adapted for building purposes, and is susceptible of that treatment which renders it fit for appropriation to exalted purposes, so the plant is material which by adaptation and modification may be used for the purpose of enhancing our pleasures.

10. The rich material furnished by the plant-form to the ornamentist is very manifest in the remains of ancient Egyptian work. Visit the British Museum, and notice the beautiful modifications and adaptations of the lotus (fig. 1), of which we have already spoken, found on the mummy-cases; or, better still, observe in the Crystal Palace at Sydenham, first the lily itself, as it appears with its blue flowers rising from the waters in the tropical end, and then pass to the beautiful Egyptian Court, and mark the manner in which it has been applied to the purposes of decoration. You may also notice in the waters the papyrus, or Scripture bulrush, and study the manner in which it has been modified by this ancient people. Make these observations attentively, and you will feel that vegetable forms, as material, were by the ancients wrought into ornament, as they are now fabricated into modern devices.

11. The suggestions afforded by plants to the ornament-producer are almost as numerous and varied as the forms, and are worthy of equal consideration with them. Some leaf-buds by their evolution set forth a diversity of character in their parts, which is welcome as it gives rise to the production of a contrast of a pleasing character. This can be illustrated

by the bud of the horse-chestnut tree, as it appears in some portion of the month of May (fig. 3), when its undivided scales, or protecting envelopes, draw back, in order that the emerald plumage of spring may appear in the form of the five or seven-fingered

Fig. 3.

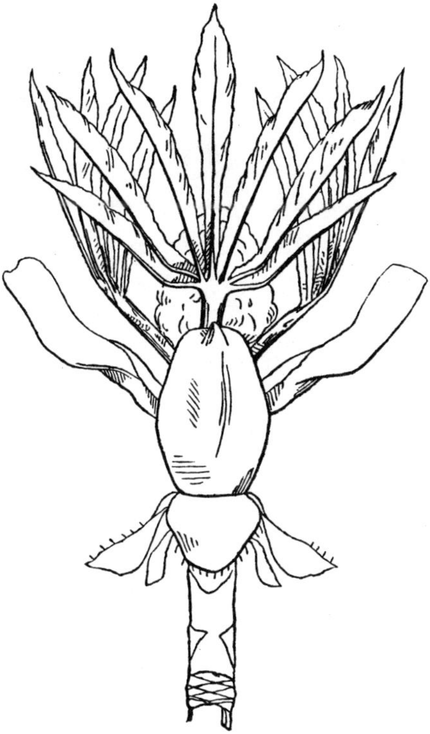

leaves of the tree. The undivided scales have a methodical arrangement, and a pleasing effect is produced by the directions which they assume; but the feature to be especially marked is the manner in which the divided or fingered fresh green leaves contrast with the whole or undivided scales.

12. A contrast not only occurs in the case of

FIG. 4.

certain buds, but is even more apparent in other instances. The botanist will observe it as brought about by exogenous and endogenous foliage; there is the grass with its narrow leaves, which are traversed by unbranched veins, and in its midst there is the buttercup and daisy, the former with divided leaves, the latter with toothed leaves, and both with leaves in which the veins have a branched and divergent disposition; and in the Tropics the large leaves of the bananas, screw-pines, and palms, with their unbranched veins, contrast strangely and pleasingly with those of the *cecropia peltata* and bread-fruit tree: the prevalence and universality of this juxtaposition of dissimilar forms will justify the statement that contrast is afforded as a primary suggestion of plants to the ornamentist (fig. 4).

13. That contrast is felt to be a want by those who are competent judges is apparent; the Greeks frequently combined the acanthus (fig. 5) with the "water-leaf," (fig. 6), the former having its foliage

Fig. 5.

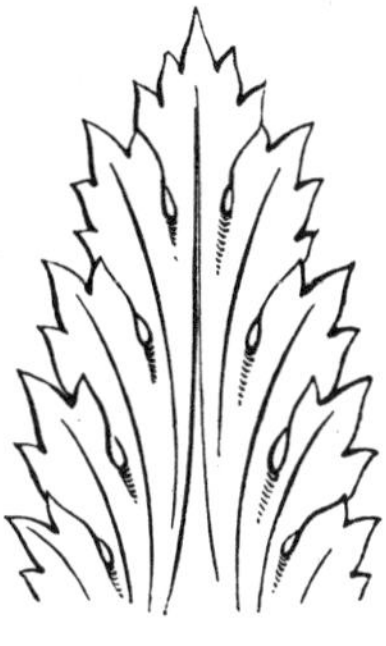

Fig. 6.

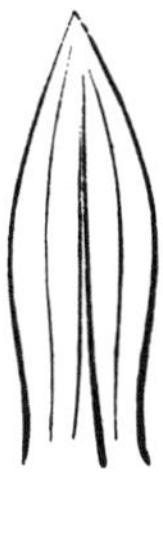

much divided, and resembling, in many respects, the leaves of the Scotch thistle, while the latter is whole, or without division, having much the form of the leaf of the lily-of-the-valley; and into the foreground of a landscape we commonly find introduced the large and divided foliage of the hogweed and the undivided swordlike leaves of the iris.

14. Suggestions of a different character are also afforded by plants. Our conception of an opening leaf-bud will be found to consist in the springing of a number of parts from the summit of a branch, some of which are reflexed to a considerable extent, while those which are internal approximate, in ratio to their position, more and more to the perpendicular; this arrangement of parts is suggested to the ornamentist, and such a conception or idea of a bud can as easily be embodied in conventional as in natural forms, for it is not dependent upon form, at least as a primary feature, but upon an arrangement of parts (Plate II., figs. 1, 2, 3, 4).

15. Vegetable nature is full of ornamental ideas: the bud, the flower, the climbing shrub and trailing plant is each an existence of which the mind takes cognizance by creating to itself internally an image of that which is external; but this image is without detail, possessing only the salient features of the object of which the conception is formed. You have probably visited the lecture-room of the Sydenham Crystal Palace, and in order to reach it you have traversed the entire length of the great building in which it is situated; but what detail have you embodied in your conception of this great edifice? You perhaps feel—and feel is the best word that we can here use—that its walls are of crystal, that the

1
2
3
4
5
6
7

centre of the building, from south to north, is a clear avenue, bounded on either side by statuary, shrubs, and courts, and that in one or two places there are water-tanks in which grow lilies and other forms of life. But have you any definite conception of the character of the plants, or of the nature of the statuary? Do you know of whom there are busts, and of whom full-length figures, or the number and order of these works of art? or are you even aware of the number and order of the larger objects, such as the courts? Have you, indeed, such a conception of a solitary vestibule that you could sketch its adornments from memory? If not, how completely is detail absent from the mental picture you have drawn of this great temple. Should we be right in surmising that your idea of this Palace is thus devoid of minutiæ, it will only be necessary, in order to awaken reminiscences of this glorious building at any future time, that we present a sketch of a crystal edifice of exalted proportions, with a promenade up its centre, and statuary, trees, and various houselike erections on either side.

16. Although such is the case, our sketch may not lead, for one moment, to the conclusion that this building has been delineated; indeed, it may be very unlike it, even so much so that it would be impossible to mistake it for this great edifice, and yet it shall call it to mind, and shall not fail in so doing in one case out of a thousand.

17. Seeing the vagueness of the mental conception, and the manner in which the mind can be induced to form to itself an image of an object by presenting to

the eye that which contains the salient features of that which the spirit is to recal, we discover the possibility of evoking ideas by the agency of decorative art such as it does not itself present, and these ideas may be pleasing or otherwise.

18. What constitutes the popular conception of a flower? Undoubtedly this—the radiation of a number of parts of the same or different shapes, and of bright colours, from a common centre. This being the case, if we place around a point a number of concentric parts of almost any shape in a radiate manner, and give to them an intense colour, we convey to the mind an idea of a flower; and by this agency a floral pattern can be constructed which in its parts presents no actual resemblance to the blossoms of any one of the eighty thousand flowering plants with which we are at present acquainted (Plate II., fig. 7).

19. By duly considering the nature of the mental conception of any flower, it is possible to suggest it in a purely ideal or conventional ornament—the fuchsia, the lily, the dahlia, and the rose, and indeed every flower can thus be suggested to the mind; but he who displays wisdom in such matters will see to it that the image suggested pleasingly or appropriately accords with the conditions which should regulate the character of an ornament.

20. While it is only necessary that a series of radiating parts be brightly coloured in order to suggest to the mind a flower, it is essential that more detail be called to aid if the intention be that of suggesting a particular flower. In this case a typical shape must not be widely departed from, and character of line,

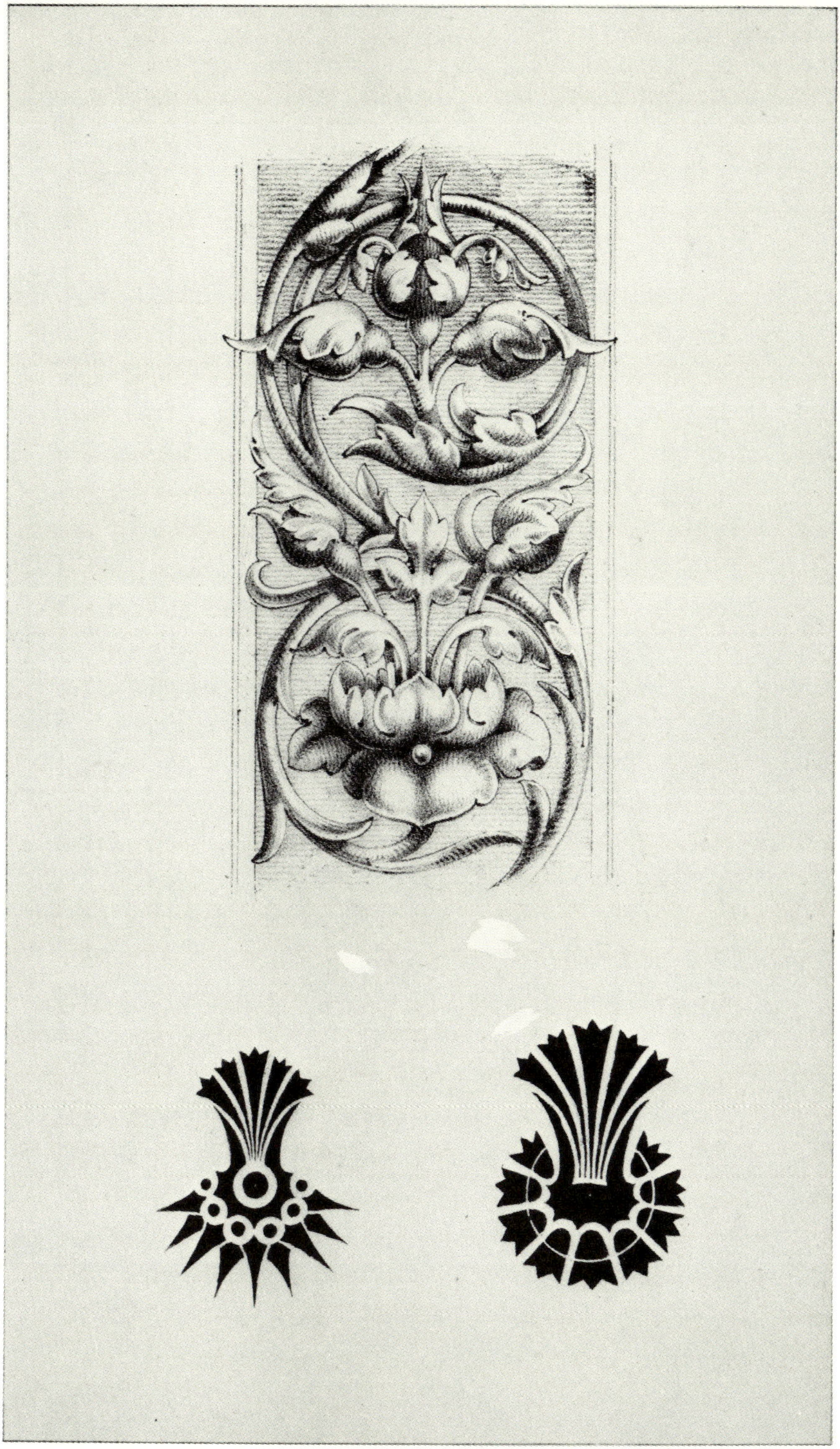

quantity, and order it may also be necessary to consider; but with consideration the plant can be suggested, and yet the ornament which suggests it be far from a mere imitation (Plate II., fig. 6, and Plate III., figs. 1, 2, and 3); it may be modified and adapted, and clothed in the mind of the artist, and yet be sufficiently characteristic to awaken the desired thought.

21. Seeing the possibility of clothing an idea in form, the plenitude of suggestions afforded by plants to the ornamentist is impressive, for every flower not only presents to us a shape, but also affords an idea which may be embodied in conventional forms, and herein rests the fulness of the ministrations of plants to ornament.

22. We shall further see the manner in which plants wait upon ornament, and the extent to which the beautiful in form is dependent upon the floral kingdom for its existence, as we proceed with our inquiry into the principles which regulate the production of decorative forms, and the few instances adduced in which plants are subservient to ornament will establish a truth which will be supported and maintained by those yet to appear.

CHAPTER III.

GRADES IN DECORATIVE ART.

1. The object we have in teaching writing in our schools is that of enabling mind to commune with mind and spirit with spirit when the bodies in which they dwell are somewhat widely separated by space or time; the secondary advantages resulting from an increase of power over the hand and a more careful mode of thought, which writing induces, may also lead us to inflict upon our children the task of learning to make and arrange characters on paper, but the primary object is that of enabling them to pass their thoughts through space, and hand them down through time, in order to instruct, gratify, or delight kindred spirits.

2. But the power to write is only of value as it enables us to embody thought in form in such a manner that others may reap from the forms the conceptions of the mind which created them; in itself the power is worthless, for no collection of indiscriminately arranged words would be of value, but as thought can be conveyed by this agency the art cannot be over-esteemed.

3. The works of Shakspeare, Macaulay, and Bacon we greatly value, but they neither command our

sympathies nor our approbation because they are combinations of characters arranged in meritorious order according to rule, but through the excellency of the mind which they make known; for while reading these works we recognise the minds of the authors conversing with our minds, as fully as if the very men were standing before us in person and speaking in audible sounds.

4. It is so with drawing, for this is but a vehicle for thought—a means to an end; and I know not that the art is worth the expenditure of labour necessary to its possession but for this one power which it has. Writing may be said to be a plain chariot in which thoughts are conveyed from individual to individual throughout a kingdom, and sometimes beyond, and decorative art an ornamental car in which mind is conveyed throughout the world.

5. One feature of especial interest, as appertaining to drawing, is this—that the language which it speaks is universal and addresses itself alike to the inhabitants of every land; then, however desirable it may be to possess the power of delineating objects for secondary purposes, the great value of such ability rests in the increased means afforded for the conveyance of thought.

6. Seeing that we value the power of drawing merely as a means and not as an end, and recognising the truth already enforced (Chap. I. s. 21), that the amount of pleasure derivable from an object is in ratio to the amount of mind which it manifests, it will not be difficult for us to determine the merits of the varied classes of decorative art.

7. Mere imitation is not ornamentation, and is no more art in the higher sense of the term than writing is in itself literature; for in order to the production of ornament there must, at least, be adaptation.

8. Our so-called "natural wall-papers" will illustrate the first, or most elementary, step taken towards the production of ornament, for adaptation has here been considered so far as is absolutely necessary in order that the design may repeat in the mechanical manner necessary to its production, but no further. Here the effort has been to imitate what is seen, and not to adapt natural forms to the purposes of decoration; the little adaptation essential is rather mourned over than gloried in, and had it not been indispensable would not have been considered.

9. If mere imitation is ornamentation, the ornamentist must at once give place to the photographer, who, by his art, repeats natural objects with infinitely more accuracy than the most careful draughtsman; but photography cannot invent, as it is devoid of the mental or imaginative faculty, for the working of the mind is essential to the production of decoration, and indeed to the creation of all exalted art.

10. We have no reason to believe that when the old Greeks were about to produce one of their inimitable statues they searched the land for the most perfect human form and simply imitated what they saw; if so, they would in all probability have made a cast of the figure. But their statues are not casts, they are types of the human shape, and have unquestionably resulted from a consideration of the form of many individuals. We imagine that prior to the

production of one of these statues an inquiry was made into the perfect form of the human body, which investigation might be long continued, and require the observation of the forms of many; and after the mind had produced to itself, in the form of a mental conception, a perfect type, worked out in all its details, an inquiry was further instituted into the modification which the form underwent as the action of the figure was brought to that required. After such preparation, the production of a figure perfect in form was comparatively easy, the difficulty rested in forming correctly the mental image, and in order to this the great knowledge was required. What we admire, then, in the Greek statue is not the manifestation of mechanical skill or labour in its construction (indeed, no thought of labour is intruded upon the mind when beholding it), but the knowledge manifested by it, and we view it with delight as the embodiment in matter of a conception of the mind.

11. We see that these works are not imitations of that which actually exists, but are mind embodied in form; and this principle applies equally to ornament, for that which is most pleasing and meritorious is that which has most fully a mental origin, while that which sets forth things seen is of a less satisfactory character.

12. Natural adaptations, we have seen, are the lowest form of ornament, but the next step, which is much more exalted, consists in the "conventional treatment" of natural forms.

13. Vegetable nature treated conventionally will not be found to be far removed from truth, but will

be merely a natural form, or a series of combined forms, neither marred by blights nor disturbed by winds, adapted to the fulfilment of a special purpose, or suited to a particular position; for the most perfect examples of what is usually termed "Conventionalized nature" are those which express the intention of nature, if we may thus speak, or are manifestations of natural objects as undisturbed by surrounding influences and unmarred by casualties.

14. In attempting the production of the ideal of nature the utmost care must be used in order to discriminate between truth and deformity; the beauty of a plant cannot be said to be repeated in a drawing if deformity is portrayed; deformity must be altogether absent from a perfect work, for art has a more exalted mission than that of setting forth what we loathe.

15. How repugnant to refined feelings it is to behold a marred feature or disfigured limb perpetuated by statuary as a work of art; and although deformity in plants may not be so manifest, and hence not so offensive, as deformity in man, yet its presence detracts from beauty, even if the source of the evil be unperceived.

16. Conventionalized plants, we say, will be found to be nature delineated in her purest or typical form, hence they are not imitations, but are the embodiment in form of a mental idea of the perfect plant; but this ideal figure is the subject of a process of adaptation.

17. The intention of the plant in producing the bud of the honeysuckle is that of developing leaves of

certain shapes in a given arrangement while they are yet varied in direction (Plate II., fig. 6); but should a leaf be partially destroyed by blight or some insect preying upon it, it will be necessary to perfect it; and should the leaves on one side of the stem be smaller than those on the other, owing to their receiving an insufficiency of light, they must be enlarged; or should any part be drawn aside from its proper position, it must again be restored to its place; and in deducing from many examples the laws and forms of nature, and producing a truly typical shape, rests the chief secret of the production of the conventional forms of plants.

18. Still following an ascending series, we find the next grade of decorative art in the embodying in form a mental idea which has been suggested by nature, and yet the form neither represents any actually existing object nor any intention of nature.

19. The illustration already given (Chap. II., s. 19) of the manner in which a mental conception can be embodied in a new form, as that of an opening bud or of a flower, will likewise tend to elucidate our present meaning, and it will be unnecessary that we spend time in showing that a composition may embody an idea derived from a natural object, and yet not imitate the object the spirit of which it embodies, for we have already seen that this is possible.

20. It is apparent that the latter class of ornament is a greater development of mind than either of those which we have already considered, for in the "Natural treatment" there is the smallest possible amount of mental power embodied, and although in "conven-

tionalized nature" there is much more, yet there is here considerably less than in the embodiment of a mere feature in ideal forms.

21. Purely ideal ornament is that which is most exalted, it being wholly a creation of the soul; it is utterly an embodiment of mind in form, or an offspring of the inner man, and its origin and nature give to it its elevated character.

22. It would be folly now to enforce the statement that conventional ornament is the highest branch of decorative art, as we have shown that the exaltation of ornament is in ratio to the amount of mind manifested in its creation; and the most careful consideration will verify this truth, for ideal creations are as far above imitations as the mind is above the body. Many brutes can imitate, but they cannot create.

23. We are now enabled to classify ornaments by affinities resulting from equality in respect to the embodiment of mental power; thus we have as examples of the first or most exalted grade of ornament the Greek, the Moorish, the early English, much of the Indian and mediæval, many features in the Japanese, and some parts of the Egyptian and Renaissance. In the second class, following a descending series, we may instance much of the Egyptian and Chinese, and a few features in the Greek and Japanese. In the third, a great portion of the middle-age work, especially the later Gothic, and many parts of the Chinese and Indian; and in the fourth, or last class, much of the Pompeian and our modern floral patterns.

24. While we have thus instanced a few examples

which seem, on the whole, illustrations of the grades to which we have called attention, yet in several cases the same part of a decoration will, by two features, illustrate two grades. Many of our floral patterns, while they belong to the fourth or lowest class in view of their composition, belong also to the first or second in view of their colouring, for it is in some parts purely ideal, and in others merely suggested by natural objects, and were it not for the more exalted character of the colouring these patterns would be much coarser than they are.

25. While there are these grades of ornament it must be remembered that there can be good and bad in each class, for if a purely mental creation has not its parts subjected to the laws or requirements of ornament, which are yet to be pointed out, it cannot be satisfying, and the pain resulting from beholding an imperfect decoration will be in ratio to the class to which it belongs, and will be greater as the class is exalted. Purely ideal art, when degraded, is most offensive, as it is the nobler part of man so debased as to be contemptible.

CHAPTER IV.

THE AFFINITY OF THE ÆSTHETIC ARTS—MUSIC, ORNAMENT, ETC.

1. We have now to bring forward the result of a series of experiments made with the view of ascertaining whether a manifest harmony exists between music and decorative art; for should a coincidence in certain particulars be established between form and sound, the rules of music might be serviceable in pointing to those which regulate the distribution and proportion of the members of an ornamental composition.

2. Our effort at reducing to rule the production of ornament, which has hitherto seemed to result from caprice, has not issued, we fear, in the discovery of principles of much importance to the decorative arts; yet as the character of the research and the results gained are alike curious, we introduce them, begging that they be received indulgently.

While we are aware that the unravelling the complexity of ideas which prevail respecting the art of decoration into a tangible code of laws has not been accomplished, we feel justified in introducing this chapter, as ornamentists are too prone to disregard kindred studies and trust to empirical knowledge.

General cultivation tends to the extension of power in a particular direction, and the laws of associated arts will frequently admit of interchange.

3. Not only do the varied branches of the fine arts abide too fully within their own immediate spheres, but art and science dwell too widely apart. We look on the one hand at the man whose refinement of mind enables him to produce lines, and forms, and combinations of the fanciful which charm all who behold them, and we look on the other hand, and see him who penetrates the deep and beautiful mysteries of scientific truth; but alas! like two links of a broken chain, they dwell apart. This chapter, and that, especially, on "The Power of Ornament to express Feelings and Ideas," will show the desirability of uniting these; and if we might become the little link which should bind these two great links together we shall indeed be glad.

4. Art and science may also be pictured as two angelic figures, entrusted with missions alike beneficent, who see in each other much that they admire, and yet there seems to be something about each which tends to repel the other. I tell you that this repulsiveness only results from their not possessing each other's more intimate acquaintance; may we, then, introduce them to each other, or rather become for the first time a priest, and bid the twain be one.

5. And our task is pleasing—for whether we become the priest or the little link, we shall be the means of uniting those which will live a long life in the enjoyment of each other's caresses, and a life which will be crowned with the happiest issue.

6. The following statements will at once set forth

the existence of a similarity in the manner in which music and decoration influence the mind. Music, by its grateful strains, lulls the spirit to sweet and joyous forgetfulness, and retards its unruly outbreakings; so does ornament, for it too can soothe, and entrance, and hush to reverie, or kindle joys.

7. Our attention is here awakened to the existence of a similarity of music and ornament in the power which they each possess of working upon the mind. But some will object to the assertion that ornament enjoys the power of working thus fully upon the spirits, but this is due to the rarity of worthy decorative examples. Some of our cathedrals awe by solemn grandeur; the Greek Court in the Crystal Palace at Sydenham speaks impressively, yet softly, of subtle refinement, and the Alhambra overpowers by lavish richness; but these Courts suffer from publicity, and the fact that when viewing them we mingle with irreverent spirits.

8. That music and decorative art are twin sisters engaged in the same beneficent work of gladdening man while on his pilgrimage through life and reviving the downcast spirit is apparent, for they both soothe and lull to reverie; they both awe by solemn grandeur and provoke joyous mirth; they both deepen sorrows and enhance joys.

9. What has been said of the power of music and decoration is also true of poetry and pictorial art, hence the æsthetic arts are intimately related, but on the affinity of poetry and pictorial art with music and ornament we must not enlarge.

10. The existence of a deeper unity between form

and sound might be expected in view of many considerations. A fiddle-bow drawn across the edge of a sheet of glass on which sand has been sprinkled, while giving rise to an audible sound, causes the particles of sand to collect into given shapes, which vary according to the note produced, and these sonorous figures are of a very interesting character. That it is the sound which brings about the arrangement of the sand particles is manifested by the figures arising from the production of the different notes being dissimilar, and by the fact that if sand is sprinkled on the tightly strained end of a drum and a powerful note is produced by a cymbal or other instrument, the sand will arrange itself into form, although removed to a considerable distance from the source of the sound.

11. Sound results from waves of air striking the drum of the ear (which is a tightly stretched membrane, resembling in many respects the end of a common drum), and imparting their vibratory motion to it, and through its agency to the nerves communicating with the brain; and the height or depth of the note is determined by the lengths of the waves of air which mediate between the sonorous body and the drum of the ear.

12. Having ascertained the lengths of the waves which give rise to the varied notes, and the fact that these bear a singular relation to each other, the propriety of inquiring whether lines of the same relative lengths (fig. 7) are in any way related suggested itself, and there seemed to be a probability in its favour, as beautiful proportion must always be

of a more or less subtle character, as we shall hereafter see.

FIG. 7.

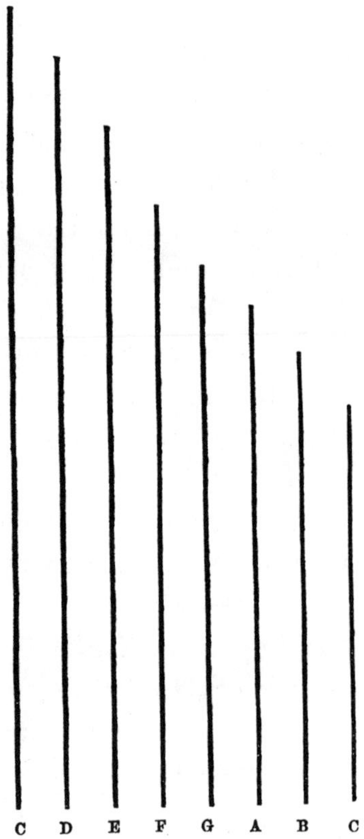

13. The form resulting from placing in a radiate manner lines which bear the same relation to one another in point of length that the waves producing the notes in the scale of music do is exceedingly pleasing, and corresponds, when repeated or doubled, with the anthemion of the Greeks (Plate IV., fig. 1); this is remarkable, that the curve formed

C
D
E
F
G
A
B
C

by the apices of the radiate members is of a high order, and one in which subtle beauty abides.

14. The next experiment, in which chords are embodied as well as the scale (which latter occupies the central portion of our diagram), and the parts spring from a spiral (Plate IV., fig. 2), is curious, inasmuch as it suggests activity or motion; it has a splashing

FIG. 8.

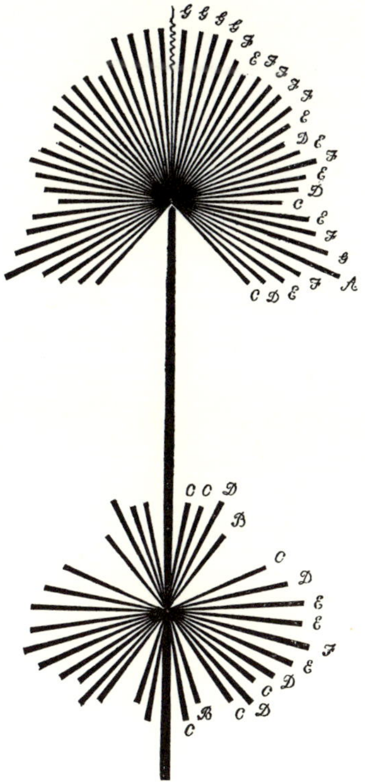

character, but here also the ornament is repeated, on the principle of the double monogram. A third experiment, in which the air "God save the Queen" is translated into form is more curious than beautiful, yet the result is not, even here, altogether devoid of interest * (fig. 8).

15. These are some few of our investigations into the extent and nature of the harmony existing between music and decorative forms, and although they cannot be said to have resulted in any great good, yet harmonious chords in music, when translated into form, furnish examples of beautiful proportion.

* For the sake of effect, the first portion of the tune occurs in the lower half of the figure.

CHAPTER V.

ANALYSIS OF ORNAMENTAL FORMS.

1. In ornament, as in science, it is necessary to have recourse to an analytical method. The deep insight into the mysteries of the atomic world with which we have been so fully favoured of late is mainly due to the energy displayed in the analysis of substances. But it is not by the chemist alone that the effort is made to ascertain the composition of bodies, for the zoologist and botanist likewise strive to observe the unit out of which all vital forms are wrought, and to discover the *ultima thule* of life. An analytical investigation is at all times interesting; the chemist, while repeatedly reducing a substance to a lower condition, is advancing towards the simplicity of the material with which nature works, and by this he learns that out of a few elementary principles "the great globe," with all its vital forms, are wrought. In like manner the botanist and zoologist discover that out of one primitive form—a little cell or

bladder-like body of extreme smallness—all conditions of life arise.

2. Pleasing as analytical research is, it has a value beyond that due to the gratification which it affords; for by a knowledge of the composition of substances, and the manner in which elements are combined into forms of utility and beauty, we alone have that power over the diversified forms of matter which enables us to mould them into shapes necessary to our well-being and comfort. No one imagines that the photographer could operate as well were he ignorant of the composition of his bath and of the nature of the substances which he employs. It is a keen perception and deep knowledge of the powers of the substances which he is using that tend most to success. And no one thinks that the artist would use his colours with the same freedom were he ignorant of the fact that green is composed of blue and yellow; our power is dependent upon the knowledge which we possess of the materials used, and he who would combine elements must be content to study the nature of that which he uses and its capacity for union.

3. The principles here set forth are fully applicable to ornament. The combination of forms cannot be carried on successfully unless the forms themselves are understood and the conditions of their association ascertained; but as the chemical student first devotes himself to analysis and only attempts the combination of elements into useful forms of matter at a later period of his career, it will be found desirable for the student of decorative art first to apply himself to analysis, and ultimately to the combination of forms.

4. In viewing an ornament, after weighing its merits and judging of its effect, the first effort should be to reduce it to its most elementary con-

Fig. 9.

dition, or most simple unit. The common geometrical form which we give in fig. 9 is merely a square with semi-circles removed from the sides and

Fig. 10.

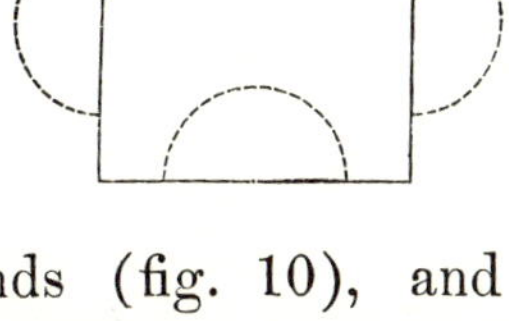

added to the ends (fig. 10), and the more complicated pattern given in fig. 11 will be seen to be the same rectangular figure with small segments of a particular form taken from definite positions and added elsewhere with the same exactness (fig. 12).

FIG. 11.

5. The practice of this principle of analysis will

FIG. 12.

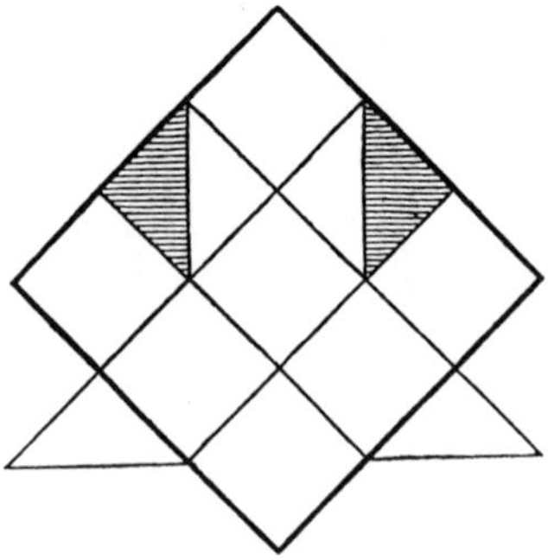

give the mind greater power over forms than can be gained in any other way. Already we see that the removing a portion of a simple form from one place

FIG. 13.

to another affords the means of bringing about an

FIG. 14.

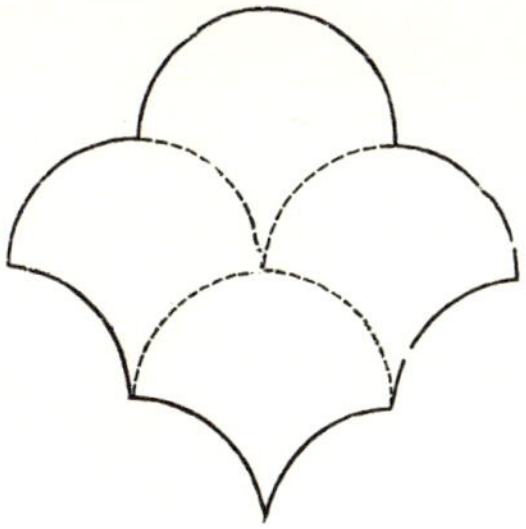

endless variety of shapes. The form given in fig. 18

FIG. 15.

is but a series of forms (fig. 14), such as are given in fig. 15.

Fig. 16.

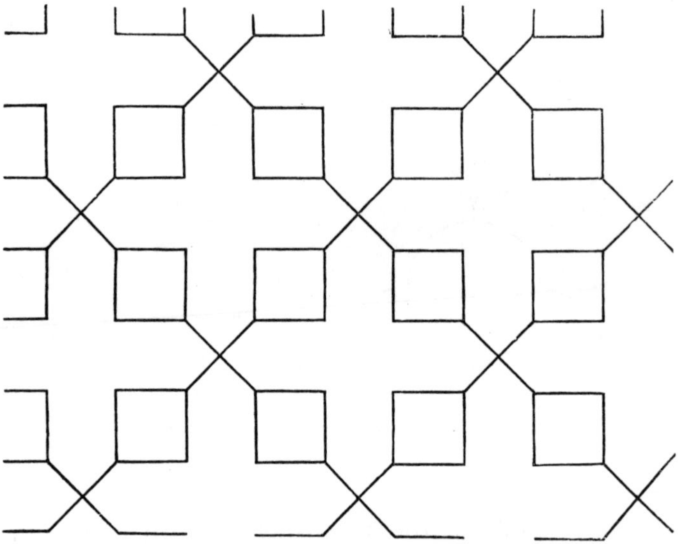

6. It is not only desirable to reduce a form to its lowest unit, but it is well to ascertain whether it can be formed of diversified elements. The form fig. 16 may be conceived to result from a combination of

Fig. 17.

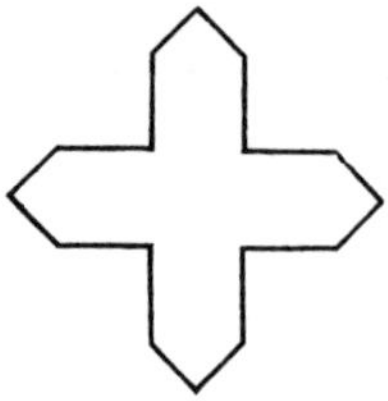

Fig. 18.

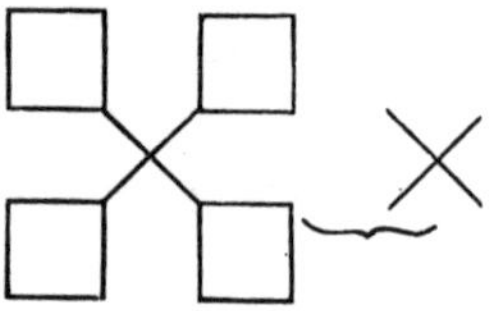

shapes such as are seen at fig 17, with spaces between them, or of the two units shown at fig 18, or of squares connected by crosses (fig. 19), or of

FIG. 19.

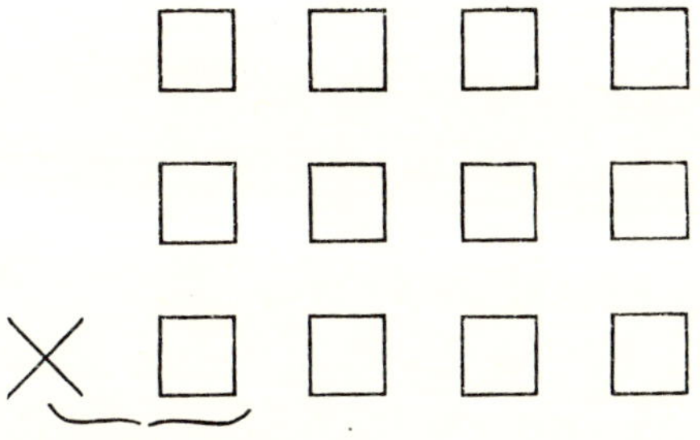

FIG. 20.

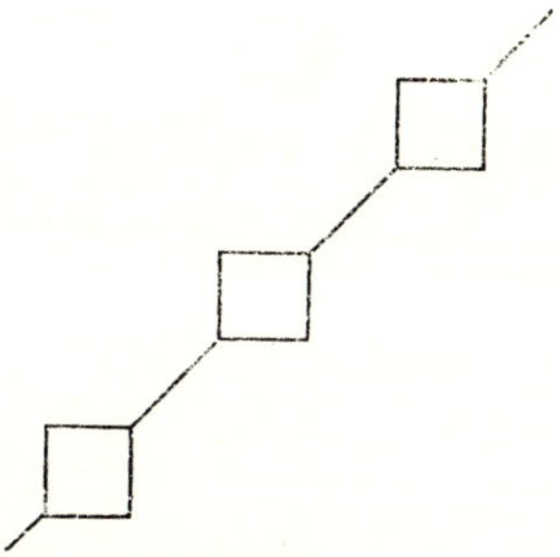

diagonal members of the character set forth by fig. 20

FIG. 21.

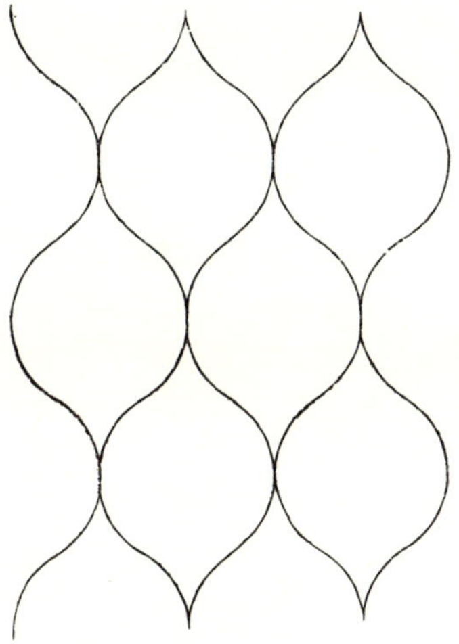

FIG. 22.

crossing. In like manner, the form given at fig. 21

FIG. 23.

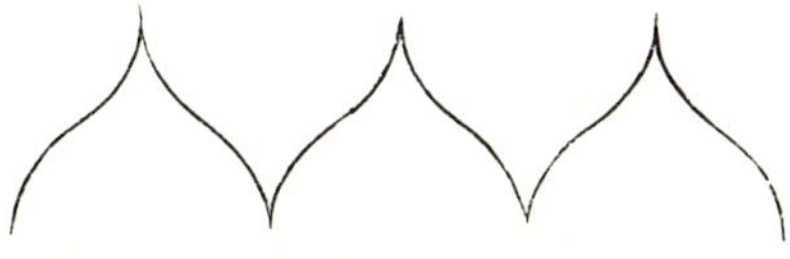

may be regarded as made up of the waved line

(fig. 22), the horizontal cuspid line (fig. 23), or of the form fig. 24, arranged closely (fig. 25), or so as to leave alternate spaces of the same form (fig. 26).

Fig. 24.

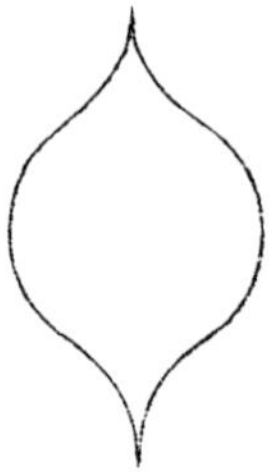

7. Simple and frivolous as this mode of investigation seems it is yet of the utmost value; for the power of combining forms, and of fitly dividing them into

Fig. 25.

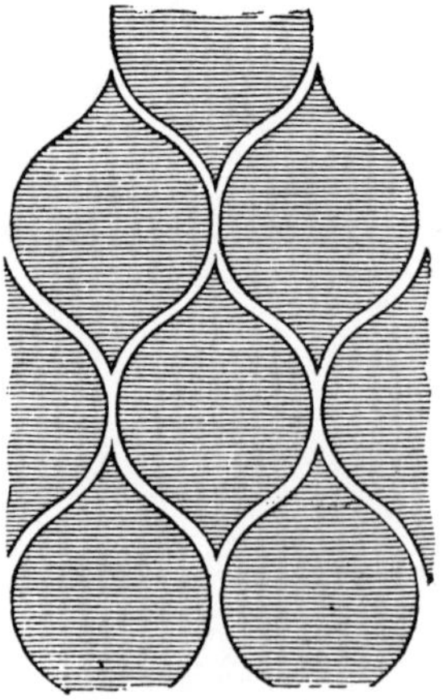

Fig. 26.

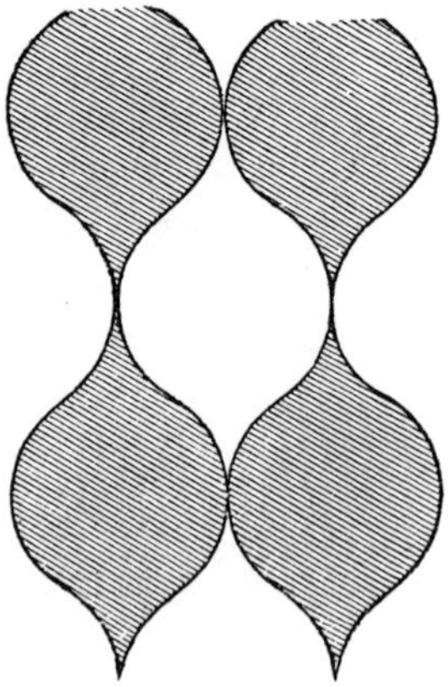

smaller parts, will be more readily gained and more fully possessed by the continual application of this mental process of analysis to the forms viewed than by any other mode of thought. But the power thus afforded for the division of forms we shall shortly have a better opportunity of showing.

Fig. 27.

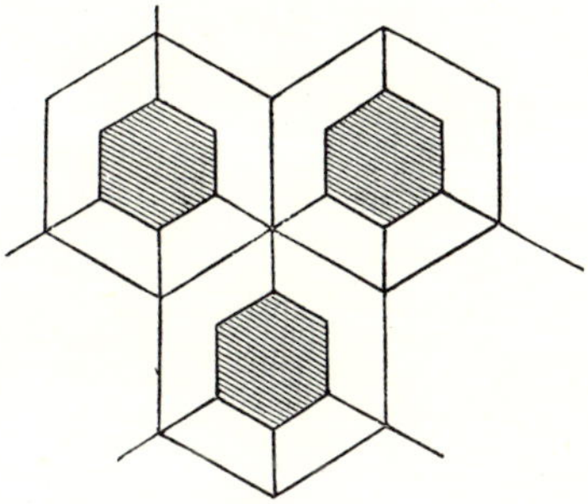

8. The forms which we here delineate (figs. 27, 28, 29, 30, 31, 32, 33, 34, 35), result from the unit

Fig. 28.

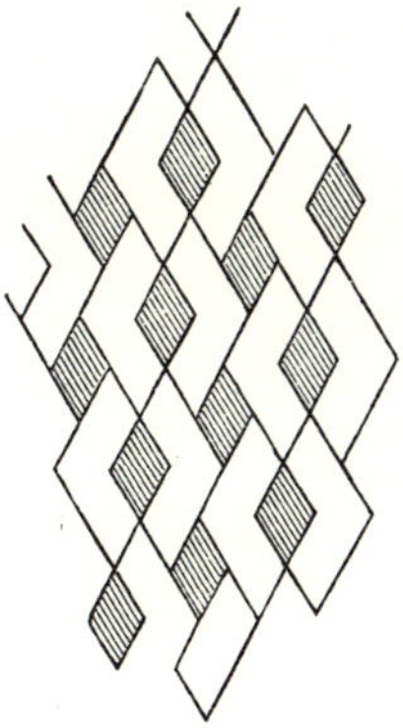

Fig. 29.

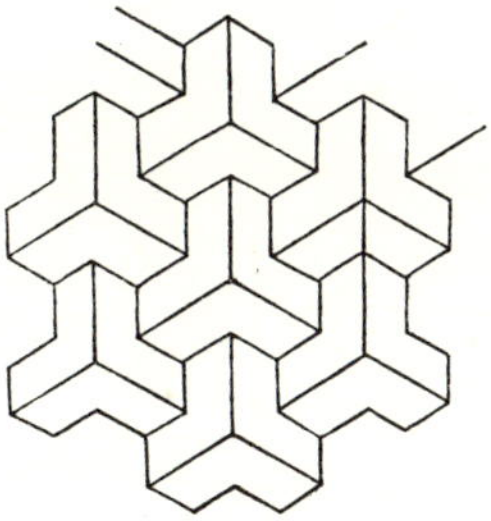

fig. 36, but this may be termed a compound unit, as it, together with figs. 37 and 38, can be reduced to the diamond, or lozenge form (fig. 39), as shown in figs. 40, 41, and 42, and the lozenge results from two equilateral triangles (fig. 43), hence it is also compound, and the triangle is the most simple element of the series of figures which we have just presented.

FIG. 30.

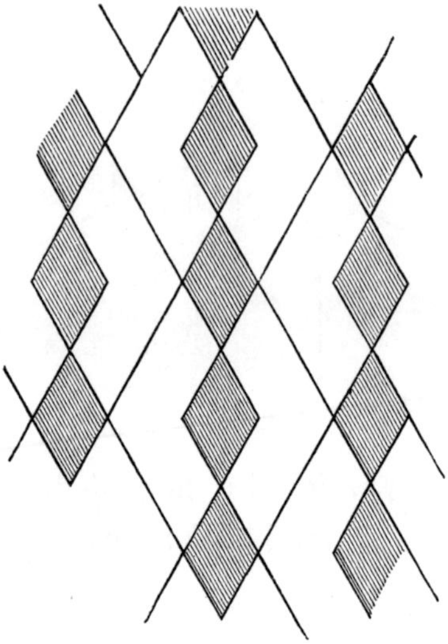

9. We are now enabled to set forth the advantage

FIG. 31.

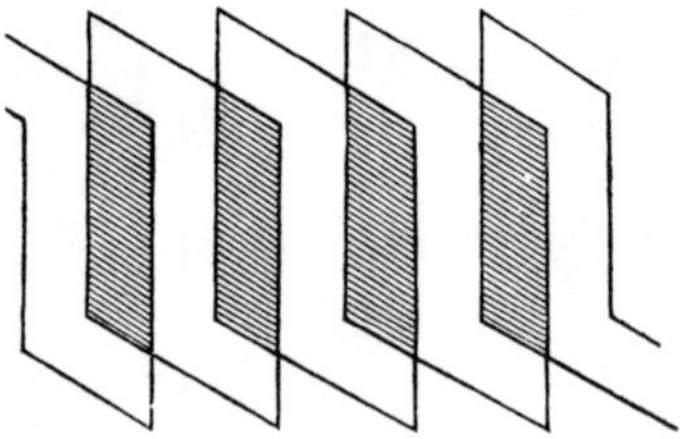

of this knowledge in the division of forms. The form

FIG. 32.

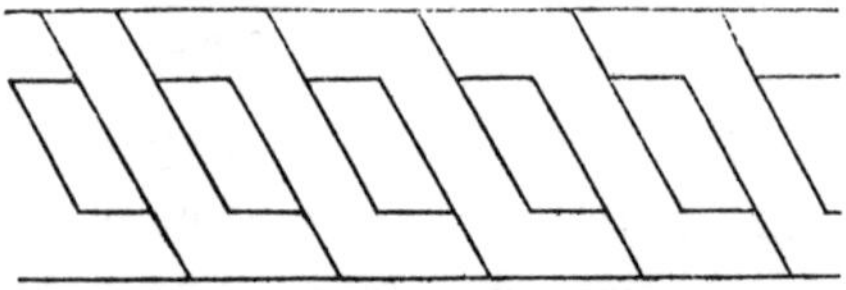

fig. 38 could not be more appropriately broken up than

by its division into the units of which it is composed,

Fig. 33.

Fig. 34.

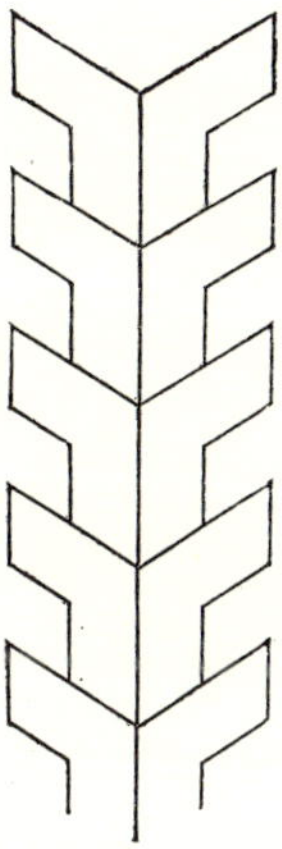

Fig. 35.

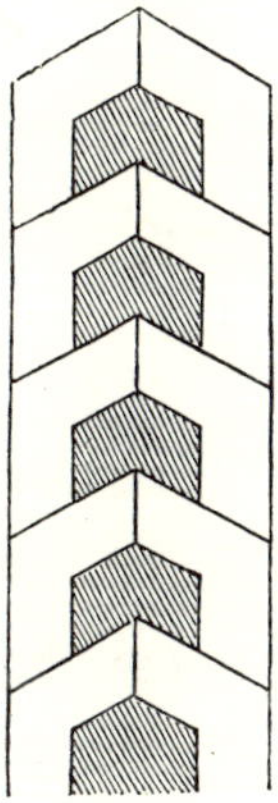

as shown in fig. 46, or into certain combinations of

Fig. 36.

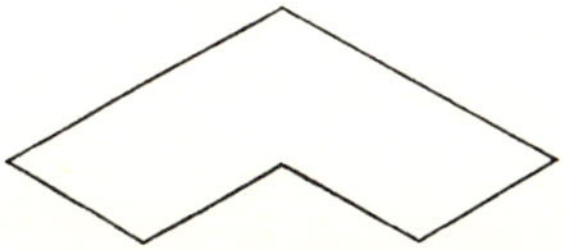

Fig. 37.

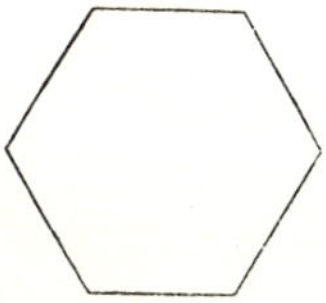

that unit (figs. 41, 44, 45); and similar instances might be multiplied to almost any extent.

Fig. 38.

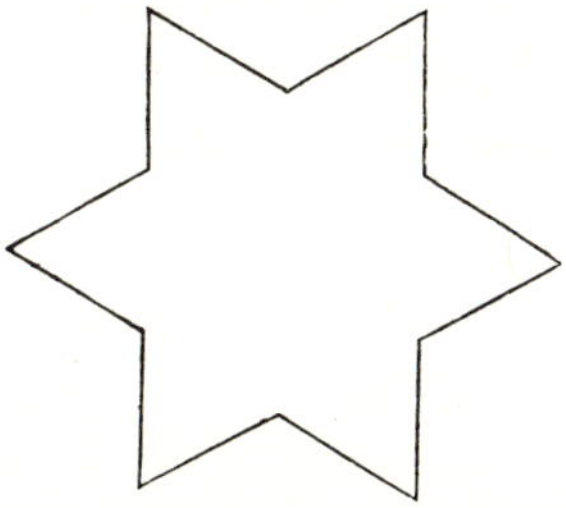

Fig. 39.

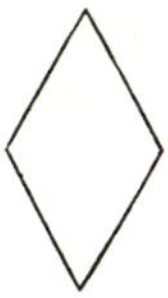

10. Fig. 38 may further be regarded as made up

of two hexagons, the one of which is divided into parts; for the six triangles which surround the hexa-

FIG. 40.

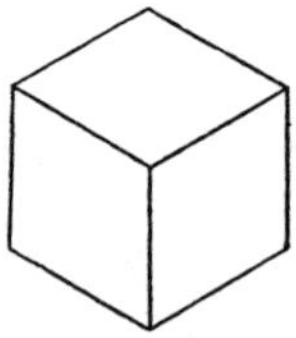

FIG. 41.

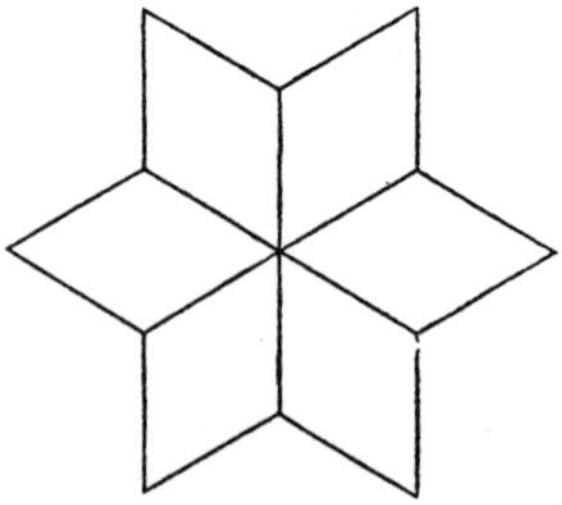

gon (see fig. 45) unitedly make up another hexagon, each triangle being the sixth portion of this figure

FIG. 42.

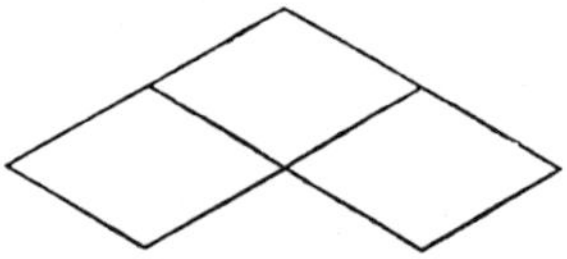

FIG. 43.

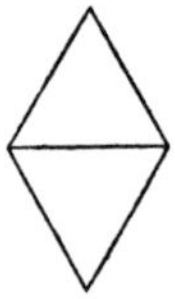

(fig. 47). This will be the better seen if we fold one member over (fig. 48). Fig. 47 may be regarded, then, as fig. 45 folded up.

11. The student will find it of the utmost value, in prosecuting these inquiries into the units of forms, to cut out a series of figures in mounting-board, and form shapes by their combination; by this he will be ascertaining the adaptation of the form for union and the unit of the forms produced at the same time. The practice afforded by the Chinese puzzle, known to most schoolboys, is good,* but is of a less useful

* This puzzle, together with a number of forms to be produced, and a key, accompanied Beeton's Annual for 1861.

character than that afforded by more regular figures, as the forms produced are in themselves valueless,

Fig. 44.

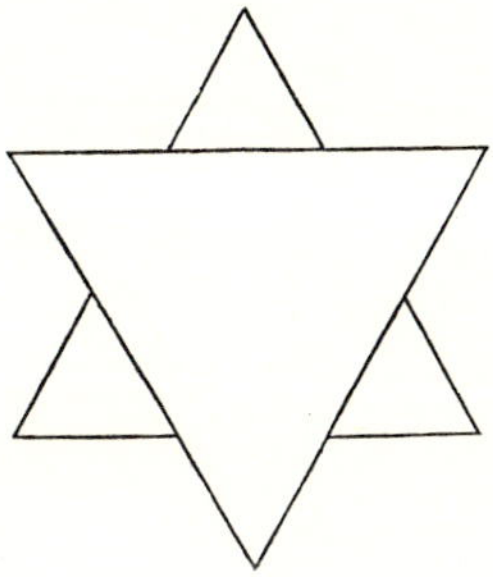

Fig. 45.

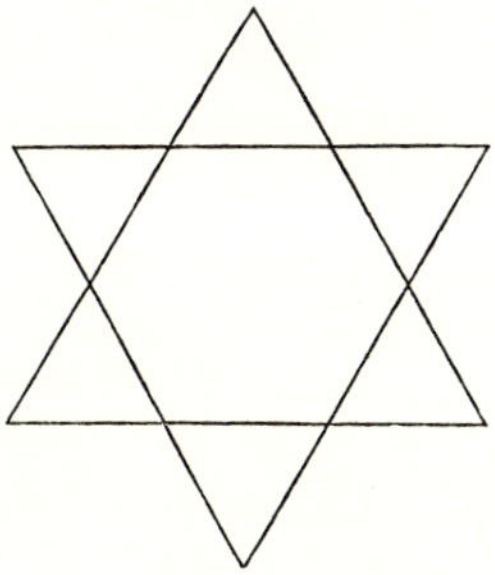

and the practice only tends to the education of the mind, while two ends might be answered at once.

Fig. 46.

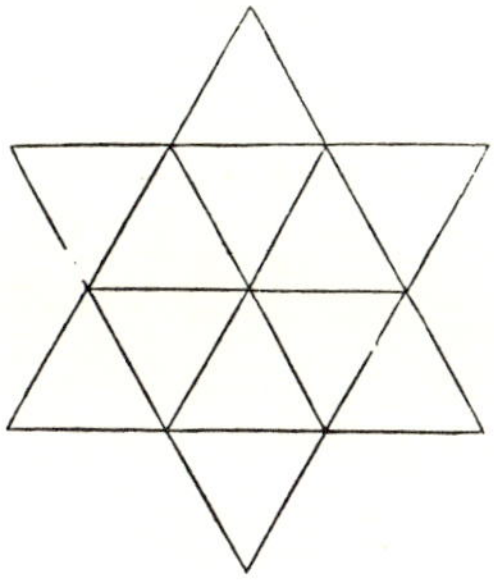

12. Seeing the value of analysis as applied to orna-

Fig. 47.

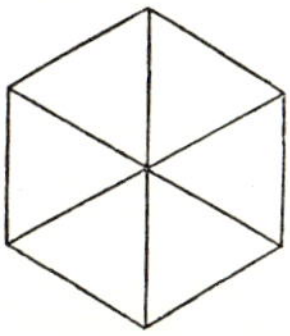

ment, it is well that we give one illustration of the

capacity of a form for entering into union. The series of forms commencing with fig. 27 have all

FIG. 48.

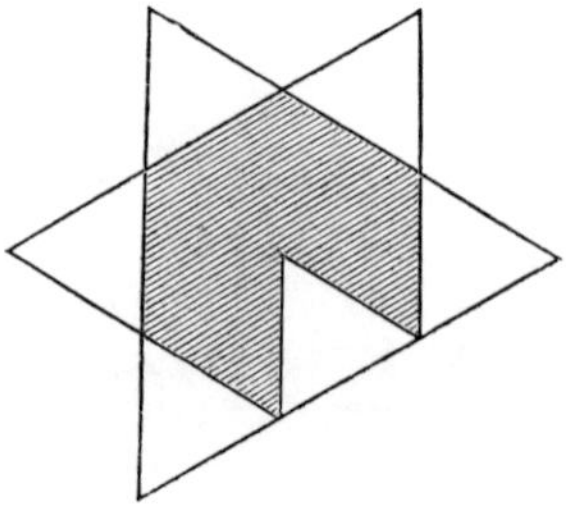

resulted from the triangle, and the majority could be formed of the compound unit, the lozenge: this latter

FIG. 49.

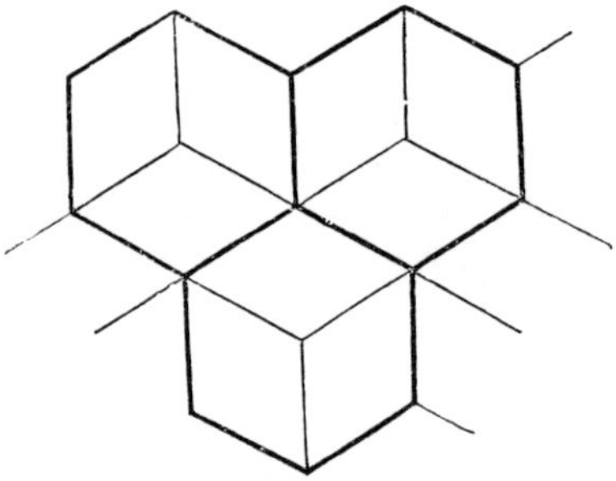

form we shall use as an illustration. Added to the figures already given, the following (figs. 49, 50, 51, 52, 53, 54, 55) will suffice for our purpose, as they represent the principal varieties of pattern which this form makes. Figs. 49 and 53 result from the form being placed together in such a manner that no ground is seen between; in the one case the lozenges are erect, while in the other they are so disposed as to form hexagons. In figs. 50, 51, 52, 54, the forms

Fig. 50.

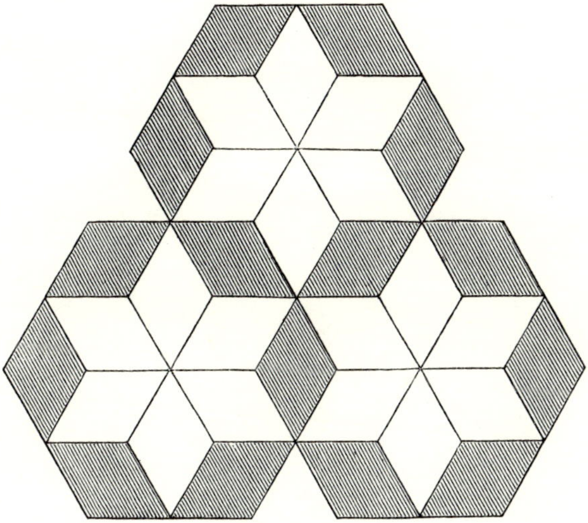

Fig. 51.

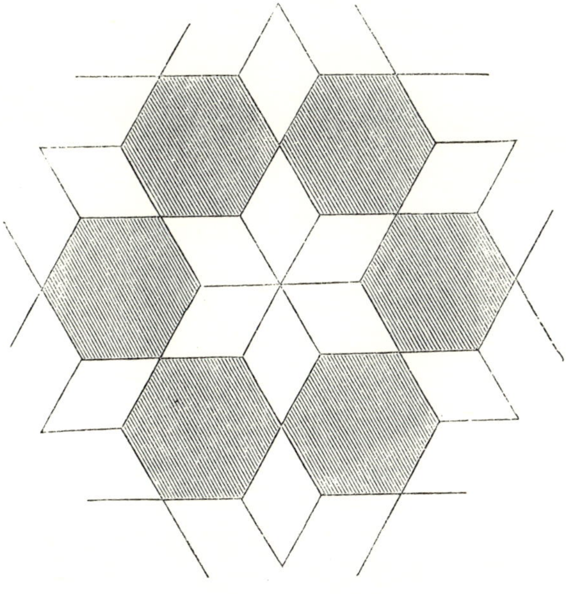

are so placed as to leave variously-formed spaces

between them; and in this way great variation in the pattern can be brought about.

FIG. 52.

13. Before dismissing the lozenge, we will treat it

FIG. 53.

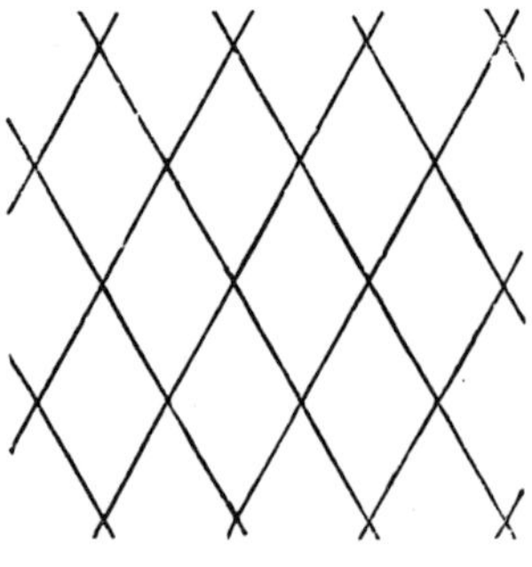

in the manner suggested in section 4; thus we cut arcs from its lower portion and add them to the upper (fig. 56). This gives us a new form capable of entering into combination (figs. 57, 58), and in the

Fig. 55.

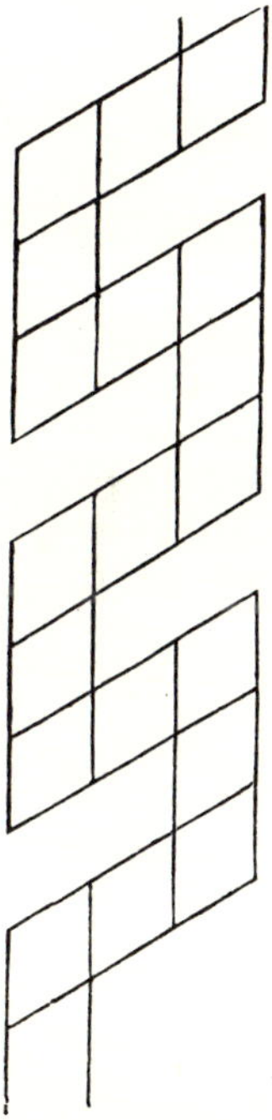

Fig. 54.

character given at fig. 57 may be classed with scale patterns.

Fig. 56.

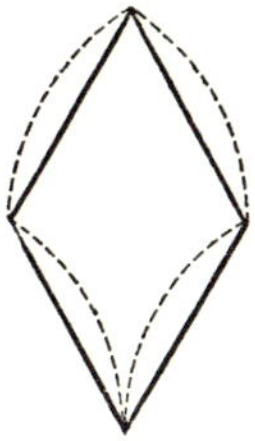

14. We have illustrated our subject exclusively with geometrical shapes, but the principle must be pursued in relation to all ornament. We have

selected geometrical forms simply because they enable us more readily to set forth what we have in view;

FIG. 57.

and all that can possibly be done, in a work like the

FIG. 58.

present, is the illustration of principles which must be pursued and worked out by the individual.

CHAPTER VI.

ORDER.

1. No one questions the loveliness of flowers; at least, such as do strangely differ from the majority of mankind; yet, incredible as it may appear, we have met with two individuals who do not esteem these charming objects. The one was a glazier, occupied in repairing one of our conservatories at the time when he uttered this incongruous remark; but the other was even a greater mistake, being a lady who not only disliked flowers but was deaf to the soft strains of music while she yet had ears, was blind to the virtues of the rougher sex while she yet had eyes, and was untouched by the soft influences of love, and yet a lady; but whether a lady-form merely, in whom nature had omitted to plant a soul, or a lady-body which had been furnished with the smallest specimen of such a commodity that was ever vouchsafed to the human frame, it is difficult to say. Happily these are exceptional examples of the race; and the class which they represent form such a minority in the great family of mankind that we need trouble ourselves little with their likings.

2. Flowers are nature's ornaments and the rich decorative forms in which she loves to array her-

self; they are as living jewels, the floral efflorescence of earth. This being the case, we may safely take them as types of ornament, and deduce from their habits and mode of growth laws which shall govern the production and arrangement of forms destined to minister to our pleasures. We may observe them attentively and study them carefully in order to possess an accurate knowledge of the character of the shapes and nature of the effects which we behold; and although the mind feels pleasure in detecting resemblances to that which it loves to contemplate, there is no occasion for the apprehension of degeneracy of taste through the observation of the majority of plants; for the forms which spring up in the garden, the shapes unfolded on the hedge-bank, the prostrate herbage of the moors, and the sturdy trees of the forest, all tell us that in the parterre, by the wayside, and in the sylvan aisles, Beauty loves to dwell and blossom forth.

3. In this, and the majority of the following chapters, we shall refer to the teachings of nature as given us in plants, in order to the illustration of the subject under consideration; but this we shall do with caution. For, while we feel that plants are the ornaments of earth, and that their forms are in the majority of instances of a beautiful character, we are also aware that it does not necessarily follow that because a thing is natural it is beautiful; for, while the braying of an ass is natural, it is yet anything but agreeable. It is so with plants; while the majority are beautiful, there are yet those which have but little of this quality.

4. We shall refer to plants, we say, for the purpose of illustration and of discovering the habits of nature in respect to the point under consideration; yet we shall regard as the greatest authority the opinions of men who have cultivated their sense and perception of the beautiful throughout a long series of years, and shall approve those practices which have received the sanction of the masters of the great art-epochs, and commend themselves to our judgment by fitness and consistency. Yet principles manifested by the great majority of plants, and especially by those which are richly endowed with ornamental beauty, we shall value as important.

5. In pursuing this conduct some will say that we are slighting nature, but this is not the case. Will any one say that the odour of the garlic is as sweet as that of the violet or rose? or as delicate as that of the lily-of-the-valley? Will any deny that the aroma of the garden balsam (*Impatiens noli-me-tangere*), crown imperial (*Fritillaria imperialis*), or of the stinking morell (*Phallus impudicus*) is less fragrant than that of the carnation or magnolia? or will any assert that the fruits of the bitter-sweet (*Solanum dulcamara*), mountain-ash (*Pyrus aucuparia*), and grape, are alike palatable? that the roots of liquorice and gentian are alike agreeable, and that wormwood and cinnamon are equally grateful to the organs of taste? To do so would be considered madness; and to assert that all plants are alike perfect in form—yea, that some are not altogether without beauty such as ministers to the gratification of the eye—is an equal folly.

6. We now proceed to consider our subject, in doing which we notice first, that in the great majority of the best ornaments of every age with which we are acquainted a principle of order is manifestly present, and this will be found to result from the decorative scheme being founded upon a geometric basis. In an ornamental composition we usually find that certain parts are emphasized and are especially attractive, these have determinate positions, even if the other parts are somewhat confused in arrangement; yet, in the best decorative compositions of all ages a methodical arrangement is found to prevail throughout the entirety of the ornament.

7. There is but a small amount of evidence that a scheme, of whatever character it is, has resulted from the operation of mind in the absence of the manifestation of order, for while confusion may result from casualties order can alone spring from intelligence. This being the case, it is manifest that the prevalence of order throughout every member of a composition, however extended, enriches it by adding to its merits, for we have already seen (Chap. I., s. 21) that the merit of an ornament is dependent upon the extent to which mind is embodied in it.

8. We now turn to the vegetable kingdom in order to discover whether a principle of order prevails throughout the structures which are here presented to us.

9. Upon beholding plants we are at first impressed with the idea that their growth is governed by no very rigid principle, for we see the convolvulus winding its way in graceful freeness around the branches of

the hawthorn bush, and the honeysuckle wandering whither it will amidst the closely-packed habitants of the thicket, and even the little daisy beneath our feet appears to protrude its leaves and blossoms as it pleases, and the stately tree surely generates its wide-spread arms just as it likes. Freedom seems the rule, and not order.

10. No habit has been attended with more injurious consequences, as respects the favourable development of ornamental art, than that of regarding vegetable structures in the cursory manner that we have done, and deducing from this hasty contemplation the broad and erroneous conclusion that plants are not developed according to fixed laws. It is this which has given rise to our so-called natural-wall-papers, and it is this which has led the great majority of the people to foster the idea that ornaments developed in obedience to law are stiff and ungraceful. The Germans are greater thinkers than we are, at least they dwell more upon the laws of nature, as a people, than we do, and there the art is more rigid in character and more obviously governed by law in its development than it is with us. The utmost care should be used before arriving at any broad and sweeping conclusion, more especially in relation to those objects which live and grow around us, as they are endowed with a power of enlarging and increasing themselves, which operations it is difficult to conceive as being accomplished in the absence of rule or law; and if law prevails in respect to growth, we are naturally led to look for order in the arrangement of the members, as the manifestation of the presence of the hidden principle.

A very cursory glance at the guelder rose will show that the leaves are, in this instance, arranged upon the stem in an orderly manner, that they grow in pairs

FIG. 59.

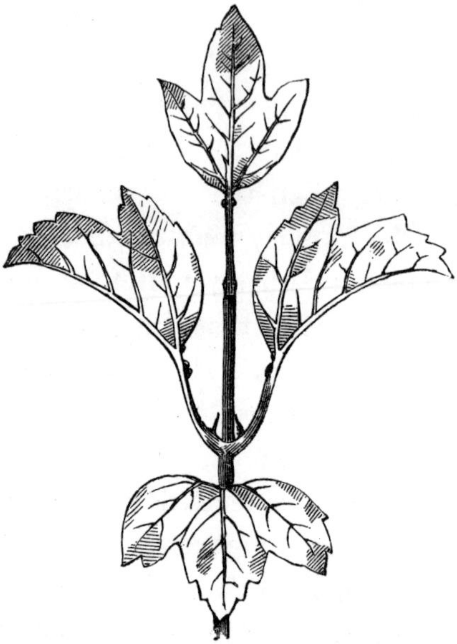

(fig. 59) which are so placed that when we look upon the top of the branch the leaves are seen to lie in four rows (fig. 60). Here two leaves are opposed to each other, one of which passes to the right and one to the left, then one of the next pair advances and one recedes, one of the next pair again passes to the right and one to the left, and so on throughout the entire length of the branch. This is not an uncommon or exceptional mode of leaf arrangement, it occurs in the garden-sage, the thyme, the white and red dead-nettle, the lavender, mimulus, pimpernel, verbena, elder, honeysuckle, lilac, sweet-william, pink, and St. John's-wort, and in many other plants

FIG. 60.

we might name. In the instances adduced it is obvious that a law of order prevails, and this will not be less apparent in the goosegrass (fig. 61),

FIG. 61.

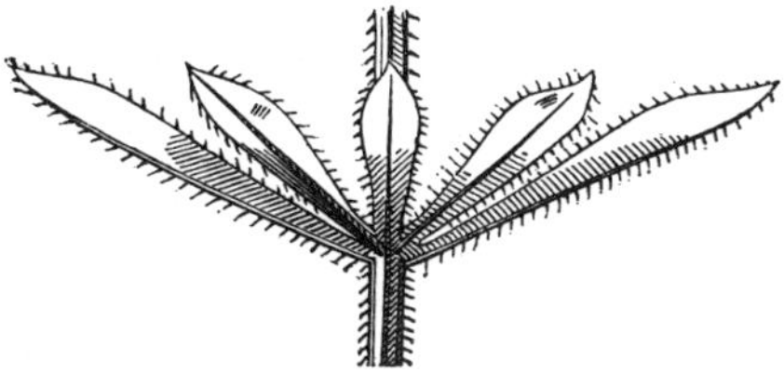

where the leaves are disposed in rings made up of about eight leaves. In the madder, rose-bay, garden-balsam, woodruff, purple loosestrife, sea-holly, and mare's-tail, the leaves are also disposed in the same verticillate or whorled manner, but the rings are made up of various numbers of leaves in the different plants, yet a most rigid principle of order prevails in each, for while the leaves are arranged in rings they are also so disposed that the leaves in the one ring fall over the spaces between the leaves in the next ring below; thus, if each ring is formed of three leaves there will yet be six rows of leaves.*

11. While the leaf arrangement is obviously orderly in the cases brought forward, it is not, at first sight, manifestly so in many instances. The leaves of the rose, apple, and holly seem to be without arrangement; but investigation has of late years done much towards the elucidation or settlement of this subject. Throughout a long period of time all leaves which were not arranged in an opposite or whorled manner were regarded as without order in their arrangement; but Bonnet noticed that, in some instances, they had a spiral disposition, or were so situated that a thread wound in a corkscrew-like manner around the stem touched the base of every leaf; but it remained for the Germans Schimper and Braun to investigate the multitude of examples, and the French brothers Bravais to pursue the subject mathematically, in order to fully clear up the whole case before us.

12. The spiral leaf-arrangement occurs in a number of modifications which become more and most complicated in character, and has its origin, or more simple development, in those instances in which the leaves are alternately at either side of the stem

* The rows of leaves are always most readily seen by looking upon the summit of the branch.

(fig. 62). In this case the leaves are in two rows situated at opposite sides of the stem (fig. 63), yet

FIG. 62.

the leaves are not opposite, for one only proceeds from the stem at the same level, and the leaves are alternately at each side of the stem, as they are con-

FIG. 64.

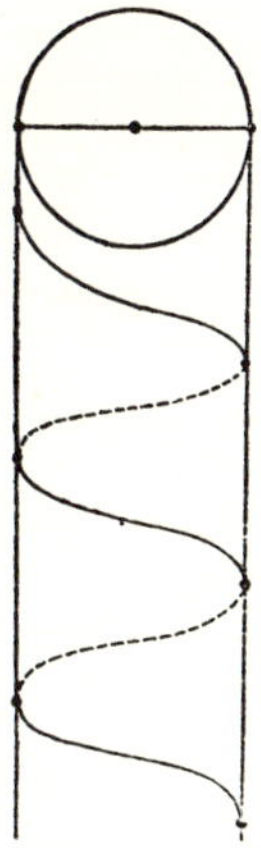

FIG. 63.

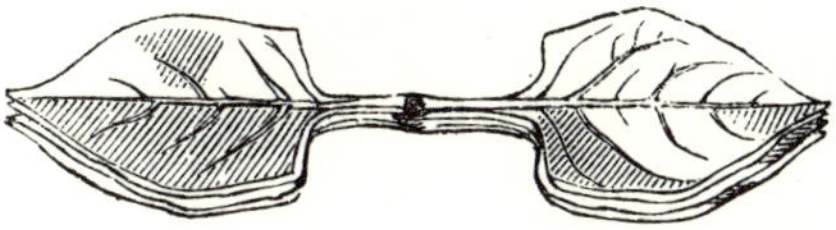

secutively higher. The spiral character of this arrangement will be better seen if we represent it in

FIG. 65.

a diagramatic manner, and indicate the positions of the leaves by dots (fig. 64); the consecutive leaves will now be seen to be equi-distant from each other; that is, leaf two will be removed from leaf one by half the circumference of the stem, leaf three from leaf two by the same distance, and so on, regardless of the extent of the series. As illustrations of this mode of arrangement we may instance the sugar-cane, leek, day-lily, traveller's tree of the tropics (*Urania*), (fig. 65), and most grasses.

13. The spiral arrangement, which we next encounter, is discoverable in the sand-grass (*Carex*), many sedges, the autumn crocus (*Colchicum*), and certain other plants, but it is usually more obscure and difficult to detect than the former (the alternate) arrangement, owing to the majority of plants which evolve their leaves on this principle having short stems. This arrangement (diagramatically delineated in fig. 66) is characterized by having three rows of

Fig. 66.

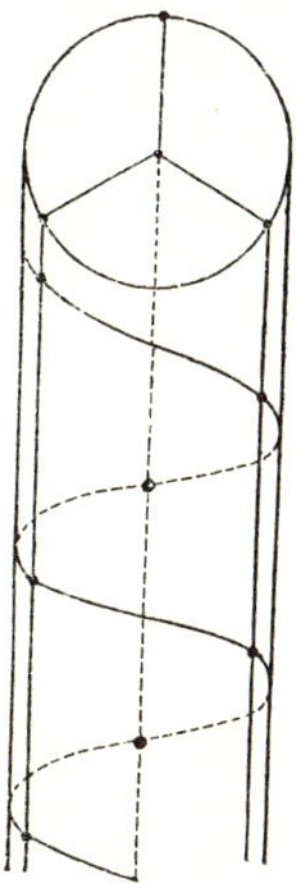

Fig. 67.

leaves, and by the leaves in the spiral series being removed from each other by one-third of the circumference of the stem, or 120°.

14. In the next advance we encounter an arrangement in which the leaves are in five rows (fig. 67), but this demands special consideration, owing to the nature of the plan here manifested. The consecutive leaves in the spiral series are here *two*-fifths of the circumference of the stem apart, or 144°. This being the case, the spiral thread passes *twice* round the stem before reaching a leaf situated over the first, while in the instances before given a leaf so situated was arrived at by making one circuit round the stem. This disposition of the leaves is of very common occurrence, being met with in the rose, apple, pear, cherry, plum, black-currant, American-currant, poplar, and oak (figs. 68, 69).

Fig. 68.

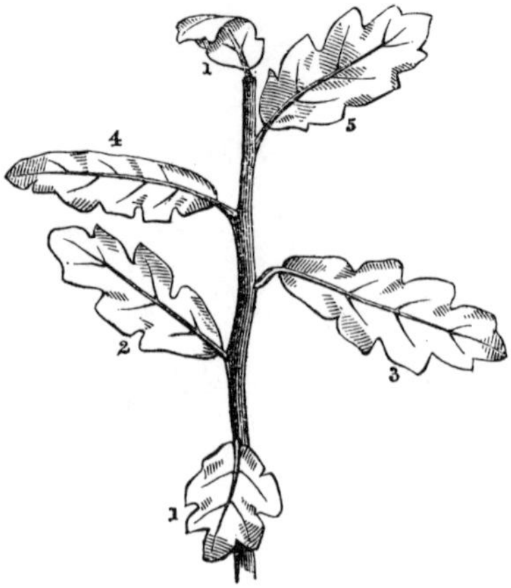

Fig. 69.

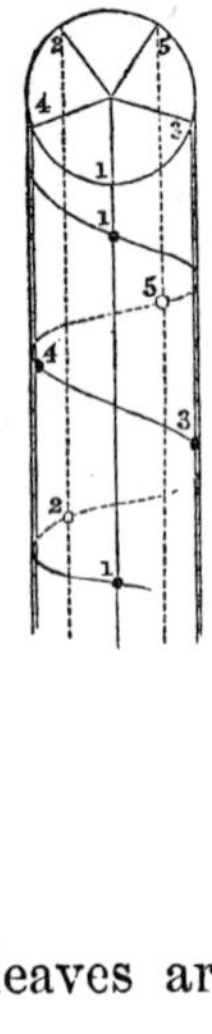

15. In the next arrangement the leaves are in

eight rows, the consecutive leaves in the spiral series are three-eighths of the circumference of the stem apart, and the spiral thread has to pass three times round the stem before encountering a leaf situated over that with which we start. This arrangement occurs in the holly, plantain, and aconite.

16. In the houseleek (fig. 70), minor convolvulus, and wormwood, we find the leaves disposed in

FIG. 70.

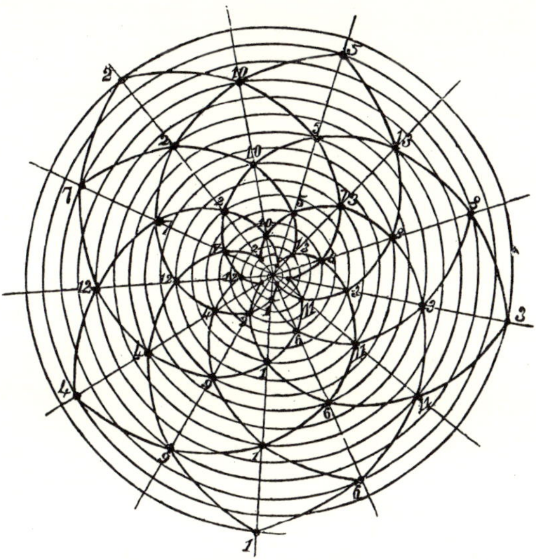

thirteen rows, the consecutive leaves removed from one another by five-thirteenths of the circumference of the circle, or 138° 24′, and that it is necessary to follow the spiral thread five times round the stem before we encounter a leaf situated over the one with which we start. These are the common modes of leaf-arrangement, yet they are not all the forms of disposition known, for a few other spiral series are occasionally met with, which it is unnecessary that we investigate, owing to their rarity.

17. We have now established the existence of order in the leaf-arrangement of plants, yet it must ever be remembered that vegetable structures are peculiarly subject to disturbing influences, owing to their not possessing the power of locomotion,* and this it is which has given rise to the popular belief that leaves are without arrangement. Blights, insects, and birds, late frosts, winds, and hail, all tend to destroy the manifestation of order in plants; and the habit, peculiar to plants, of elongating in the direction from whence they receive their greatest supply of air and light, tends strongly to disturb the natural dispositions of the parts, yet none of these influences interfere with the existence of an orderly law. Because an arm has been amputated, or the foot of the Chinese girl confined in an iron shoe, we do not say that the human figure is unsymmetrical or that the feet are cramped as the result of the intention of nature; neither can we justly say that because the development of a plant has been disturbed or marred by external influences that it is not developed according to a principle of order.

18. That leaves have an orderly arrangement is apparent, but these are but the summer garb of the plant. In the winter the branches are studded with buds, which form the winter clothing of the plant. Have these an orderly disposition? and are branches generated in order? and are the flowers thus subject

* There are a few exceptions to this in the case of the spores of the lower plants, especially Algæ, which have the power of moving from place to place.

in their development to rule? are not these parts rather protruded casually and in no definite arrangement? The answer to these inquiries will be found in the following facts. The bud normally grows immediately above the leaf, or in the angle formed by the leaf-stalk and the upper portion of the stem (fig 71), hence whatever is the arrangement of the

Fig. 71.

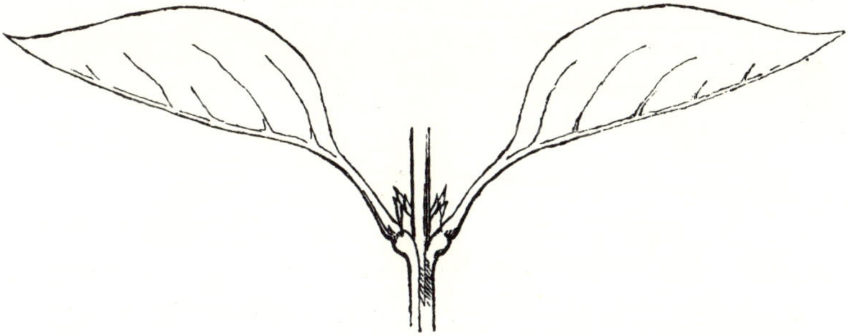

leaf such is also the arrangement of the bud.* Buds, by evolution, elongation, or unfolding, become branches, they are, indeed, only short branches waiting

Fig. 72.

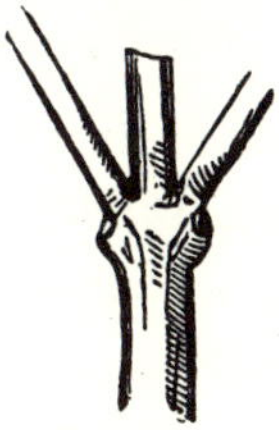

to elongate under favourable conditions (fig. 72), consequently whatever is the arrangement of the bud,

* Buds appear casually from any points of an old stem when the chief branches have been blown from the tree by a tempest, or have been removed by man, but this is only the effort made by all living objects to retain life.

such is also the arrangement of the branch.* And flowers are but stunted branches with leaves, therefore whatever is the arrangement of branches and leaves, such is also the disposition of flowers and of the floral parts (fig. 73).

FIG. 73.

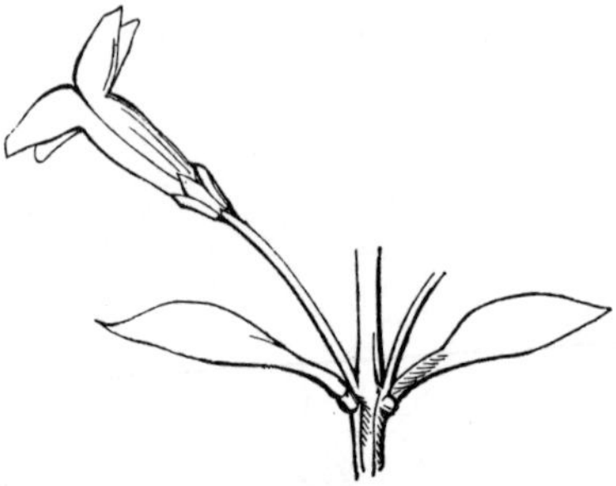

19. We now see that the varied parts of the plant are subject to an orderly principle of development; indeed, it is manifest that all the parts are thus protruded in fixed stations, for the only two typical organs of the plant are the leaf and stem, and these in their modifications give rise to all the members of the vegetable structure. The *stem portion* divides into the *branches* and *flower-stalks*, and is continued downwards as the *root*, and the *leaf type* appears as the *scales* or winter clothing of the leaf bud, the *sepal*, *petal*, *stamen*, and *carpel* of the flower, in other words, as all the floral members save the flower-stalk, and as the *fruit*. We have here not only a manifestation of the truth that a principle of order prevails in the development of plants, but also a revelation of the fact that plants are founded upon a geometric basis. The top

* It frequently happens that birds eat buds, and thus prevent a symmetrical development of the branches.

1

2

3

views of branches will make this very apparent. The opposite arrangement is associated with the square or quadrangular form, or cross (fig. 60); the verticillate, with three leaves in each ring, with the hexagon, the spiral arrangement with five rows, the pentagon (fig. 67), while the spiral disposition with thirteen rows, as manifested in the various species of house-leek, is of a complicated and beautiful character. Still, while we say that every flowering plant is founded upon a geometric basis, we know that this is frequently to an extent concealed by the slender character of the stems, and the feeble nature of the leaves, yet it is still to be found when sought. As then all plants of a highly organized character, including those which are most esteemed on account of their beauty, are built upon a geometric plan, as the most worthy ornaments have a like basis, and as order most clearly manifests the operation of mind, we deduce that order is essential to the production of exalted ornament. Yet, as in plants the geometric principle is in some instances much more manifest than in others, so in ornament we may use a positive geometric form (Plate XXVII., fig. 1), or merely indicate its presence by an orderly arrangement (Plate XV., fig. 2). It is, however, possible to give a geometric form while such does not exist separately from the ornament, as will be seen in Plate V., figs. 1, 2, 3.

CHAPTER VII.

REPETITION.

1. REPETITION is so obviously a principle of plant growth, and so manifestly a source of much of the pleasure which we derive from beholding the vegetable structure, that it will be unnecessary that we enter into any argument in order to show its value, or that we do more than point out the manner in which it occurs in the plant.

2. The first form of the plant, whatever be its ultimate character, is that of a little bladderlike body, or cell, of a more or less spherical form, and of an elastic nature. No sooner does growth commence than this little cell is multiplied, that is, it generates others similar to itself, which aggregate around it till a spherical mass is formed. Cells are now formed chiefly in an upward and downward direction, and thus that elongation in contrary courses (at the summit of the stem and point of the root) is established which characterizes the growth of most vegetable structures. At the very commencement of plant-growth repetition occurs, for the cell which at first represented the plant has simply multiplied, and growth, we see, is dependent upon repetition.

3. In order to point out the manner in which plants

1
2
3
4
5
6
7
8

further repeat themselves during the process of enlargement, we may notice a spray of any plant, that given in page 75 (fig. 62) will fully answer our purpose; a cursory glance will show that we may cut the stem through, in this instance, between every second leaf, and thus divide it into similar parts, indeed we may further divide it between every leaf and reduce it to like portions, for it is only the arrangement which gives one a direction to the right and the other to the left, that in any way necessitates a division between every second leaf in order that the branch be reduced to units (see also Plate VI., fig. 8). We have been enabled to divide the spray with alternate leaves into a number of similar parts, in like manner we can divide the stem with opposite leaves (Plate VI., fig. 7) into units which exactly correspond with each other by cutting through the stem between each pair of leaves; and the whorled leaves (fig. 61) can be sundered in a similar way. It is obvious, from what has been said in Chap. VI., s. 14, that a branch with leaves arranged in the more complicated spiral dispositions consists of similar parts repeated, for the spiral series occurring in the oak, apple, and rose repeats in every five leaves—hence if the stem is cut through immediately above the fifth, tenth, and fifteenth leaf, and so on, it will be reduced to similar parts. But each of these units consisting of a portion of stem and five leaves may be further divided, for division may take place between each leaf, when it will be found that the units (each being composed of one leaf and the small portion of stalk to which it is attached) are similar, and that the compound unit (the five leaves with their common stalk) results from the manner in which these unifoliar portions are arranged together, or the direction which each has.

4. In order that we see the manner in which repetition occurs in the formation of the branch we may have reference to its development. A little eminence appears, which is its first representative; this becomes slightly elongated, and a little protuberance appears upon its side which developes into a leaf, thus the first unit is formed; the little stem-portion further elongates and originates a second leaf, it further elongates and produces a third, then another unit is formed, consisting of stem and leaf, and however long the branch may ultimately be, its growth consists in the continued production of similar parts.

5. Repetition, manifested in a manner somewhat different to that occurring in the instances already adduced, occurs in the case of the flower. Here we have an organ formed of one or two members repeated in a circular arrangement (fig. 74), hence in this case

FIG. 74.

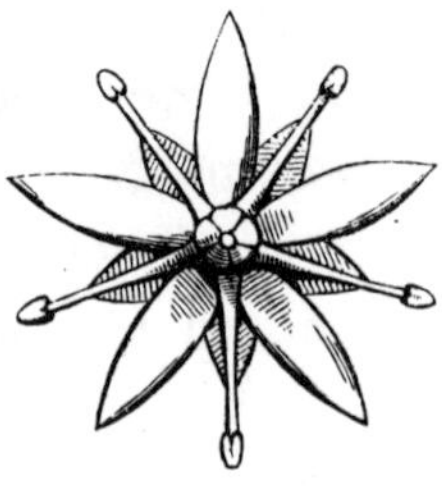

we have radiating repetition as distinguished from the elongated repetition already considered. Radiating repetition not only occurs in the case of the flower, but is seen in the top view of every branch. In that of the most simple spiral arrangement (Chap. VI., s. 12) the leaf is repeated at either side of the stem, in the opposite arrangement (Chap. VI., s. 10) in four directions from a centre, and in the spiral, found in the oak (Chap. VI., s. 14), in five directions—thus in the case of the branch a top view gives radiating repetition, and a side view elongated repetition.

6. That repetition is a primary feature of plant growth is still further shown by the branching of the plant, for when this operation takes place the plant is multiplied, and thus a repetition is brought about of that which exists. This will be seen in the case of a young plant which is possessed of but few buds, indeed of but two, one at either side. Here when the buds develop each becomes elongated into a branch, and each also sends down a portion of woody matter during its elongation equivalent to the root of the parent plant, thus the mother structure is repeated. The mistletoe, many seaweeds, and some chickweeds, as well as other plants, branch in this bifurcated manner, but in whatever way the branching takes place the repetition of the plant is brought about by this operation.

7. We have observed repetition in the plant as bringing about an elongated series and as producing a radiate series, but there are yet other forms in which this principle is manifested. We have what we term extended repetition, which is illustrated by the plants to which we have just referred—*i.e.*, the mistletoe, chickweed, and certain seaweeds, as the species of Ceramium—the ceramium ciliatum for example—

and by the compound umbel (the branching flower-head) of the parsnip, carrot, and hemlock. This class of repetition occurs in such a manner that while the plant is multiplying itself in an onward direction, it is also spreading in a divergent manner, and the individual branch is not extended in a continuous manner, as in the instances first noticed.

8. There are yet other modes of repetition found in the plant; the one is, the repetition of a spot or of a part. If we look upon the primrose from above, each flower is as a star, and the flowers but repeat the one form. This is the case in every flower-head, and in some instances groups of flowers are repeated. What is commonly called the flower of the cineraria and of the china-aster is a head made up of scores of diminutive blossoms, and yet each head of flowers is repeated in the manner of the solitary blossom in the case of the primrose. The plant is also studded in the grass as a star pattern. The plantains, which frequently have their leaves pressed on the ground, set this forth very manifestly. The other form of repetition found in the plant is what

FIG. 75.

we may term repetition with variation. This is well illustrated by the petal of the mignonette, which has the form represented in fig. 75. Here we have a number of club-shaped parts constructed on one general plan, yet all differing in form; there is a repetition, yet in the repetition there is variation. This principle is further set forth by the opening leaf-bud (Pl. II. figs. 1, 2, 3, 4, & 6); by the compound leaf, as that of the horse-chestnut; and by the total branches of a tree. We have seen that when a stem branches it simply repeats itself, yet the branch when one year old is only like the parent when of the same age. In the case of the tree with its branches, while the uppermost branches are one year old those which are lower are two, those which are still lower are three, those lower, four, and so on; and the older arms are longer than the younger; hence we have this repetition of branches, all constructed on the same type and yet all varying in length, and consequently in bulk and curvature. There is also that form of repetition corresponding with the side view of a row of trees.

9. These are the forms of repetition occurring in the plant; let us see if they have their analogues in ornament. The first form which we noticed, that resulting in the production of an elongated series, occurs in many modifications in the diversified decorative schemes; it would result from the repetition of fig. 1, Plate III., and occurs in Plate VI., fig. 6.

10. The radiating repetition has its counterpart in ornament in all those compositions which spring from a common centre, and consist of similar portions. Figs. 76 and 77, and Plate VI., figs. 1, 2, 3, 4, and 5, are examples of this form of pattern.

Fig. 76.

Fig. 77.

1

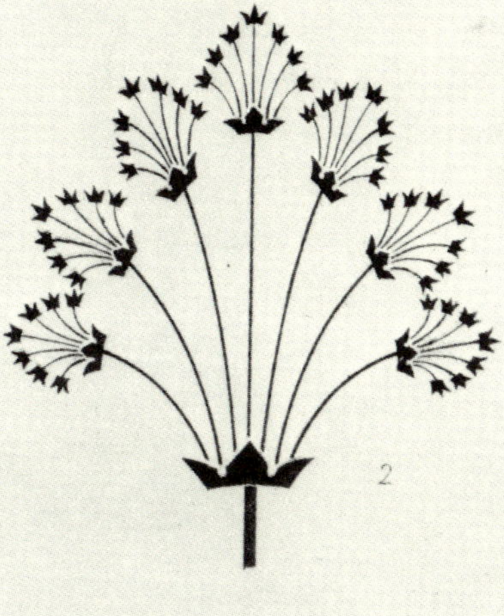

2

1
2

1
2
3
4
5
6
7
8

11. Extended repetition is not of such common occurrence as the forms just noticed, yet it has occurred in most styles of ornament, and was introduced as early as the Egyptian period. • It is set forth ornamentally in Plate VII., fig. 2.

12. The repetition of a spot, or of a part, is exceedingly common, and gives rise to a class of pattern termed a powdering: when combined with order, it has its most simple form in the repetition on a geometric basis of a dot. Repetition, corresponding with that of the leaf on the tree, appears in ornament in the manner set forth at Plate VIII., fig. 2, and that analogous to the repetition of the leaf and flower on the plant in Plate VIII., fig. 1.*

13. Of all the forms of repetition seen in the plant, that with variation is, perhaps, the most interesting; but whether this be so or not, this is certain, that some of the choicest ornaments result from the application of this principle; and it is also of interest to notice that this mode of repetition has appeared in every style of ornamentation. In the Egyptian decoration it occurred in the form set forth in Plate IX., fig. 1; in the Greek in the Anthemion (Plate IX., fig. 2), one of the most common ornaments of this period; in the Alhambra we find it in the form given in fig. 1, being the same as that of the Egyptian; in the early English as seen at fig. 3; in the Chinese as fig. 6; in the Japanese as figs. 5, 7, 8; and in the Renaissance as fig. 4. And upon this one principle an infinity of ornaments can be constructed, a few varieties of

* These two figures have been derived from the beautifully illustrated volume, by W. and C. Audsley, "The Sermon on the Mount."

which we give in Plates X., XI., and VII., fig. 1. The illustrations given, and the fact that this principle has been adopted in all ages will establish its importance and commend it to the consideration of the art-student.

14. The form of repetition corresponding with the side view of the row of trees is set forth in figs. 78, 79.

FIG. 78.

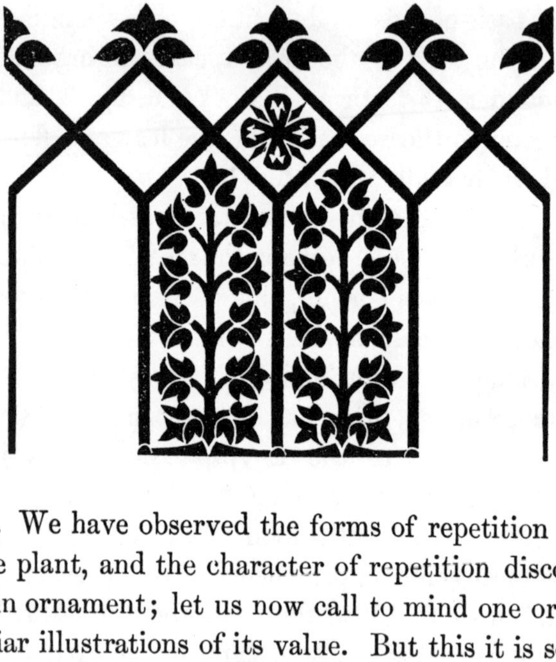

15. We have observed the forms of repetition seen in the plant, and the character of repetition discoverable in ornament; let us now call to mind one or two familiar illustrations of its value. But this it is somewhat difficult to consider apart from order; but in conjunction with order its value is very manifest. A word, if doubled, whatever it be, will be pleasing (figs. 80, 81), and yet the forms employed are not of a very ornamental character; and even a word written and folded over while the ink is wet, and

1
2
3
4
5
6

1
2
3
4
5

rubbed so that it is blotted and yet repeated, gives

FIG. 79.

rise to a blotch which is more or less agreeable. The kaleidoscope furnishes an excellent illustration

FIG. 80.

FIG. 81.

of the value of repetition. Here a few broken, irregular-shaped, variously-coloured fragments of glass become extremely pleasing when repeated; and yet the same fragments arranged in the same manner would be altogether without interest in the absence of repetition; hence it is apparent that the repetition, in an orderly manner, gives rise to the pleasing aspect. While we argue that the forms seen in the kaleidoscope are pleasing, we by no means commend them as ornaments; for, as we have before seen, unless mind is embodied in the composition it cannot be satisfying; and he who has recourse to such an agency for decorative schemes thereby manifests the poverty of his inventive faculty, and reduces himself to a mere mechanical operator.

CHAPTER VIII.

CURVES.

1. MUCH might be said concerning sublimity of line and the subtlety of curves of an extremely interesting nature, but in a work like the present, which purports only to lay down the principles of ornament, and not to furnish essays on the varied subjects associated therewith, all that can be done is to call attention to those points which are essential to good ornament in the briefest manner possible.

2. With curves it will be found that refinement rests in subtlety, for that line the constructive origin of which it is most difficult to detect is found to be the most beautiful. Thus an arc (fig. 82) is the least beautiful of curves, for its origin is instantly detected. A portion of the bounding-line of an ellipse (fig. 83) is more beautiful as a curve than the arc, for its origin is less apparent, it being struck from two centres. The curve which bounds the egg-shape is more subtle than the elliptic curve, for it is struck from three centres, hence it is more beautiful still; the curve bounding the cardioid is more beautiful still, as it is struck from four centres; and so on, in ratio to the number of centres employed in the construction of a curve, and its consequent subtlety, is its beauty.

3. We have now given a clue to the whole secret

of beauty of line; but while we say that the beauty of the curve increases with the complexity of its origin,

Fig. 82. Fig. 83. Fig. 84. Fig. 85.

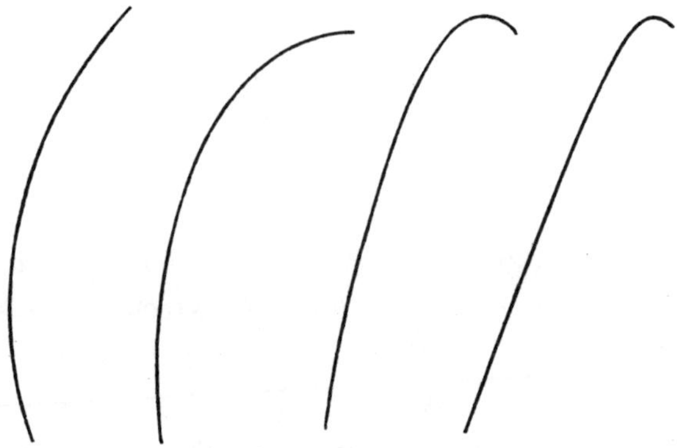

Arc. Elliptic Curve. Parabola. Hyperbola.

we make no reference to the forms which the curves bound. The curve of the circle is of a low character as a curve, yet the circle, if not beautiful, is grand, and the ellipse is, as a form, very pleasing; what we now speak of is the curve as a line, and not as the boundary of a form.

4. The curves used by the Greeks, who understood more of refinement of line than any other people, appear chiefly to be those derived from conic sections, as the parabolic (fig. 84, manifested in projecting bodies), hyperbolic (fig. 85, manifested by the capillary attraction of water confined between two pieces of converging glass), and the elliptic (fig. 83) curves, but these are not of equal merit in point of beauty, as will appear from what has already been said—the hyperbolic being the most exalted in character.

5. There are a number of other curves of a beautiful character which might be named: the catenary, or

chain curve, manifested when one end of a chain is fixed and the other is raised to any height—the involute, manifested in the unwinding a thread from a reel—the convolute or rolled-up curve of the watch-spring. But we do not wish to go into curves mathematically; we have but to deal with beauty of line, and must hence refrain from dealing with the mathematical view of the case. The secondary curves formed by the complicated modes of leaf-arrangement are of a very high order, as will be seen by a glance at fig. 70; we do not refer to the great generative or watch-spring-like spiral of this figure, but to the more rapid curves which we have drawn through the dots indicating the positions of the leaves.

6. Curves must not only be of a subtle character, but must present an amount of variety; thus that line is least satisfactory which has its halves alike; and in this respect, as in relation to the origin of the line, that which is richest in variation is most beautiful. Hogarth's "line of beauty" is well known; it is a curve of a subtle character, and presents that variation in its parts which is pleasing; but whether it is really entitled to be termed *the* line of beauty is questioned by some. Our respected friend Mr. Arthur, the drawing-master to the Prince of Wales and the Royal Family, is inclined to the belief that the pastoral crook is the true line of beauty, and this conviction is borne out by many circumstances, for this line occurs more frequently than any other in the purest early English ornaments (Plate IX., fig. 3), and is that which forms the boundary of the members of the best examples of the Greek anthemion

(Plate IX., fig. 2), hence in this conjecture there appears to be much truth.

7. There is not only beauty, but there is also power of line. A curve while yet beautiful may be feeble, or it may be full of energy and life; and that line will be found to be most powerful in character which approaches nearest to the right line.

8. The *line of life* is worthy of special consideration. Nothing is so pleasing as the manifestation of a strong vital power, and the thought of such can readily be called forth by a certain class of line.

9. The shooting forth of the buds in spring must be studied in relation to this line, and the direction which every leaf takes when given off by the stem in its activity of energetic growth. But in order that it be fully comprehended, we must have reference to the tropical vegetation, where the leaves are of colossal size and the vitality of plants is at its maximum; but as a tropical forest is not accessible to all, we give a group of young palms, in which the greatest power is manifested in the direction of the members, and in which we truly have *the line of life*.* (Fig. 86.)

10. We see, then, that if we desire to invest our ornaments with the beauty of life it is necessary that we have energetic curves, and avoid those which are feeble through being bent about like worms in agony. It is always better to approach the straight line in the curve than to err in an opposite direction.

* These have been derived from the Java views, to which we have already made reference, by Messrs. Negretti and Zambra of Regent Street and Hatton Garden.

Fig. 86.

11. This knowledge of lines will give a power which could not otherwise be possessed, for it affords a means of manifesting strength of varying intensity in conflict with power or force. A grass stem when bent by the breeze has a curve varying with its thickness and rigidity as well as with the pressure of the wind; and to a great extent a truthful idea of the pressure in the one case and the strength in the other can be given by the line.

12. If the student desires to refine his taste he must have curves of acknowledged beauty to contemplate and dwell upon; let the hyperbolic curve be formed by water between sheets of glass, and a small chain be so suspended that one end is higher than the other; let, also, the secondary curves in diagrams of the character of fig. 70 be had constantly in sight, and by thus having lines of beauty ever before the eye the feeling for refinement will be developed and enlarged, and thus the mind will be rendered more fully susceptible of detecting and appreciating beauty.

13. In relation to music, a discovery has recently been made by a Parisian philosopher, which is of inestimable value; it is a method of tuning instruments by sight with absolute exactness, so that to a standard thus afforded the most refined ear may yet be cultivated. And in relation to form, the having before the eye lines of such a character that the taste may ever be cultivated to them as standards, answers the same end as having the exact notes in music.

14. The union of curves with curves, and of curves with right lines must always take place in a tangen-

tial manner. Nothing can be more offensive than seeing one line unite with another in such a manner that it looks as if it would run through that with which it unites, or as if a gimlet-hole had been bored in the one, and the other was inserted into the opening thus formed. There must be a graceful flowing of line out of line, however far the composition may be extended—and this can only be accomplished by a tangential union.

CHAPTER IX.

PROPORTION.

1. "If nature has made the human body so that the different members of it are measures of the whole, so that the ancients have, with great propriety, determined that in all perfect works each part should be some aliquot part of the whole, Proportion is that agreeable harmony between the several parts of a building which is the result of a just and regular agreement of them with each other,—the height to the width, this to the length, and each of these to the whole." Proportion, then, consists in the combination of a unit into quantities which, when set forth by surfaces or solids, agreeably accord with each other, and tend by their relations to gratify the beholder.

2. That proportion which is most pleasing will be found to be of a subtle character; four to four is bad; four to three or five to eight is better; for it is more difficult to detect, and in ratio to the subtlety of the relationship, and the consequent difficulty of detecting it, is the beauty of the proportion.

3. In the middle-age architecture we constantly have the proportion of five to eight, which in the windows occurs in two relations (figs. 87, 88); but never find the proportion of four to four, which is inva-

riably bad; yet even a more subtle proportion than that of three to eight will be found to be most agreeable.

FIG. 87.

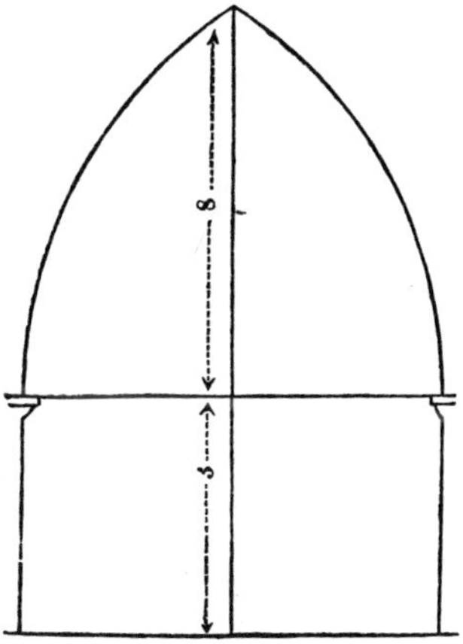

FIG. 88.

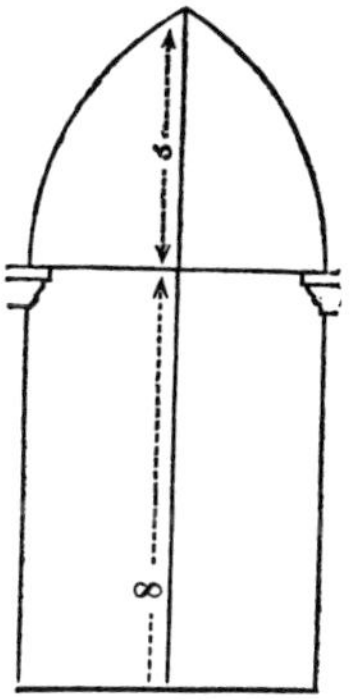

4. Professor Zeising has proposed a proportion which he terms the "golden cut," as being that which prevails throughout the entire economy of the universe; but while we cannot go with him in all the steps in his argument, we yet highly appreciate his "golden cut," and value it much as a subtle and valuable proportion.

5. In order to find the "golden cut," erect, at the end of a horizontal line (*A B*, fig. 89), a perpendicular line (*A C*), half the length of the horizontal line (*A B*); connect the remote extremities of the horizontal line (*A B*) and the perpendicular line (*A C*) by a diagonal line (*C B*). With the upper extremity of the vertical line (*A C*) as a centre set-off, upon the diagonal line (*C B*), a distance (*C a*) corresponding with the height of the perpendicular line (*C A*).

Transfer the distance corresponding with the remaining portion (a B) of the diagonal line (C B) to the horizontal line (A B), when the "golden cut" of the horizontal line (A B) is marked (by the point a'). The proportion which the smaller portion of the horizontal line (the portion A a') bears to the larger portion (a' B) is unquestionably just and satisfying,

FIG. 89.

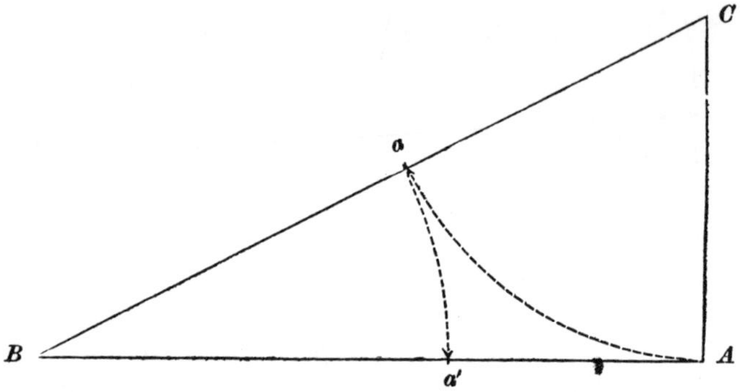

and is doubtless the best proportion with which we are acquainted.

6. Translating this proportion into mathematical representatives, the relation which 618·0339887 bears to 1000·0000000 will be found to be its representative. But to set this relationship forth in a longer series we give the following table from Dr. Zeising's work:—

1000·0000000
618·0339887
381·9660113
236·0679774
145·8980339
90·1699435
55·7280904

34·4418531
21·2862373
13·1556158
8·1306215
5·0249943
3·1056272
1·9193671
1·1862601
0.7331070 and so on.

7. In relation to the circle, the proportions given in the following figures (90) are from the same author,

Fig. 90.

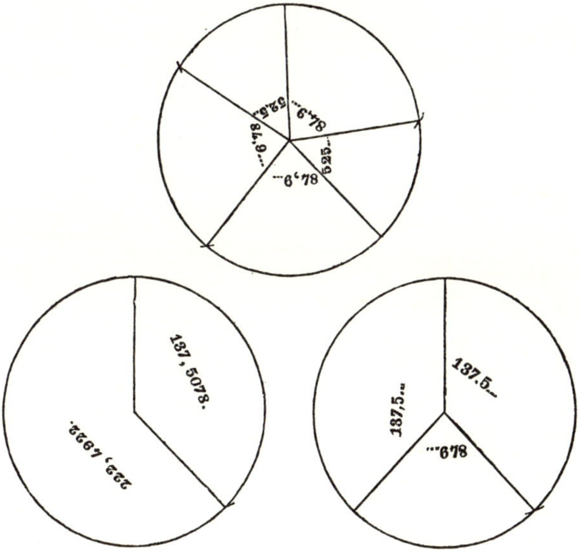

as is also the mathematical setting forth of the matter given in the following table:—

360·0000 degrees
222·4922 ,,

137·5078	degrees.
84·9844	,,
52·5234	,,
32·4610	,,
20·0624	,,
12·3986	,,
7·6678	,,
4·7308	,,
2·9368	,,
1·7940	,, and so on.

While it is unnecessary for the purposes of decoration that recourse be had in all cases to this method of ascertaining proportions, a consideration of the principle is by no means useless; but it is attended with much benefit. The mind, by contemplating those relations which are found to be most agreeable, becomes thereby cultivated, and refined, and quickened in its power of perception; hence such considerations are valuable. But while it is not necessary that we in all cases find the exact golden cut, it is essential that proportion be of a subtle character if beauty is desired.

CHAPTER X.

ALTERNATION.

1. THE term "alternation" as here used must not be understood in its common sense as synonymous with reciprocal, but as a term derived from botanical phraseology, where it has reference to an arrangement almost invariably found in the parts of the flower, and in many instances in the distribution of the leaves upon the stem also.

2. The flower is made up of four series of parts—a ring of outer leaves, which are usually green; a ring of coloured leaves; a number of awl-shaped or thread-like members terminated in knobs which are usually yellow; and a central organ consisting of one or several portions. In their arrangement these parts set forth the principle of alternation, for the members of the second series do not fall over the parts of the first series (the petals do not fall over the sepals), but fall over the spaces between them, or *alternate* with them. The parts of the third series do not fall over the constituent members of the second, but between them—they alternate with them; and so with the inner series. This condition of things is set forth in fig. 91, where the outer series of parts (the sepals unitedly constituting the calyx) is marked by

diagonal lines; the second series (the corolla) is made up of three leaf-like members; the third (the andræ-

FIG. 91.

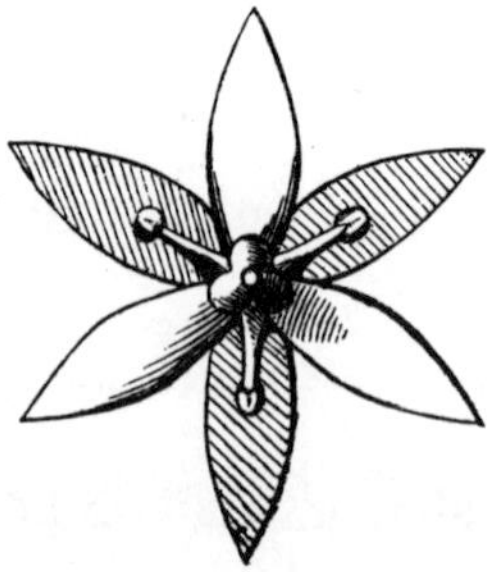

ceum) of three awl-shaped members (each a stamen) terminated in knobs, while the central consists of but one three-lobed portion, and the consecutive series alternate with each other as we have seen. (See also fig. 74.)

3. Not only is the principle of alternation manifested in the disposition of the floral parts, but, as we have already said, it also occurs in the arrangement of leaves upon the stem.

4. In the case of alternate leaves (fig. 62), the second leaf does not fall over the first, but is situated at the opposite side of the stem, and in opposite leaves (fig. 59), the second pair does not fall over the first, but over the spaces between the units of the first pair, as is fully seen in the top view (fig. 60); but in the verticillate or ringed-leaf arrangement (fig. 61) the analogy with the disposition of the floral parts is most fully seen, for here the parts of the second ring do not fall over the members composing the first ring, but between them.

5. Alternation is a principle, we see, manifested in

plants, and it is a principle which also prevails in ornament, and appears to be called for by a natural desire; for an observation of figs. 92 and 93, will at once show that that (fig. 92) in which the principle of

FIG. 92.

FIG. 93.

alternation occurs is more satisfactory than that in which the principle is not acknowledged, through the smaller leaves being hidden by the larger. So manifestly is the one superior to the other, and so intuitively is this superiority recognised, that we are half disposed to regard it as a principle called for by

FIG. 94.

FIG. 95.

instinct. Yet the desire for an acknowledgment of this principle may arise from our intimacy with flowers, in which this is almost an invariable law.

6. The two simple forms set forth in figs. 94, 95,

derived from a warming-pan of early date, are of interest in relation to this subject, as this principle of alternation is in them fully acknowledged, and the whole of the ornamental character which they manifest (it matters not how little this be) is due to the acknowledgment of the law of alternation which they set forth. This principle is also recognised in Plate XII., fig. 1, an ornament of a somewhat different character.

7. The principle of alternation is not of modern introduction into ornament, but has been recognised from the earliest times. In the Egyptian ornaments it is fully set forth, as a glance at the ornaments of that period will show, and in the decorations of the Indians (Plate XII., figs. 2 and 3), Greeks, and Romans it also occurs. The Greeks frequently formed mouldings and borders of leaves (fig. 96); but there was not merely those forming the external series, but others *alternating* with them, and the echinus (fig. 97), or egg-and-tongue moulding of the Romans,

Fig. 96.

Fig. 97.

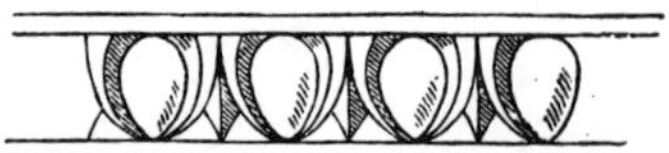

consists of oviform masses arranged in a horizontal series, with tongues or arrow-heads between them *alternating* with them. Thus that form of alternation which has its analogue in the side view of the foliaceous branch is set forth in ornament; while the principle, as manifested in the top view of the leaves and in the flower, is seen in nearly all the radiating ornaments of the different art-epochs. (Fig. 98.)

PL. XII.

8. The principle of alternation is not only of interest as it occurs in the ornaments of every age and in

Fig. 98.

plants in the manner which we have already seen, but also because it occurs in diversified forms, or under dissimilar aspects in the vegetable structure, and because there seems to be a special object in its adoption.

9. Relative to its purpose it seems to be designed as a means of calling attention from a union of lines which in all cases is unsatisfactory. Thus in fig. 93 the eye immediately rests on the spaces between the four members, and is thus conducted to the junction of lines, while the alternating parts give an outward direction and thus direct the attention from the point of weakness; and whether the alternation be manifested in the circular ornament or the elongated series, the same end is answered.

10. In plants we have a continual confluence of lines; leaves are given out from the branches in all directions, and branches from stem, but the branching of the stem is concealed by the foliage during the summer months, in which period alone plants present themselves in an ornamental aspect; and while the leaves are given out in countless numbers, the means devised for calling attention from the union of the leaf-stalk with the stem are endless. The bud arises in the angle formed by the upper surface of the leaf-stalk and the stem (fig. 71), and thus acts as the alternating members in the examples already adduced; but a special provision for so arresting the attention that it shall not fix upon this confluence of lines is also made, as a pair of small leafy or mem-

Fig. 99.

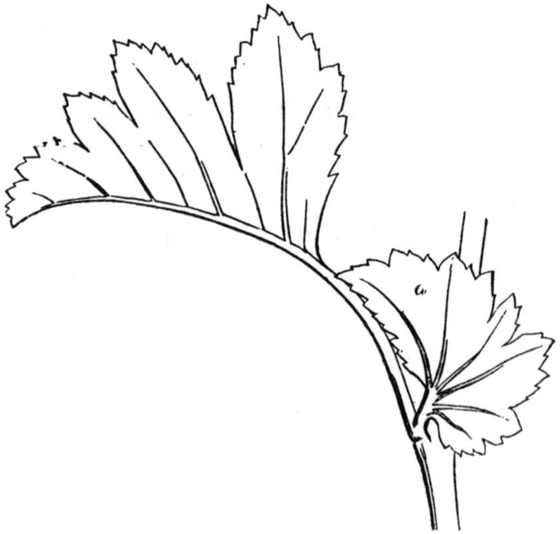

braneous organs (stipules), of a form rich in subtle beauty, conceal this juncture, or so arrest the attention that it does not fix upon it (figs. 99, *a*, and 100)—thus the principle of alternation is substituted by the growth of special organs. And there seems a con-

FIG. 100.

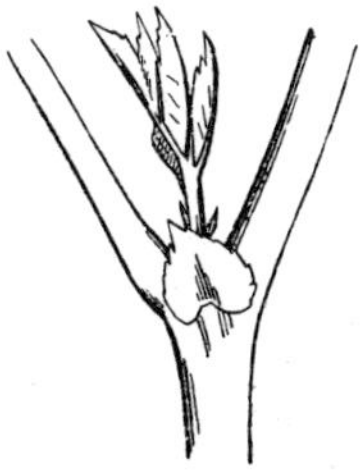

sistency about this principle of alternation when carefully examined which is not apparent when but a cursory glance is given, as will appear upon our making one or two observations. Whatever be the form of each unit of the outer series of parts in the flower, as in fig. 92, it has practically, when seen from above, the contour of the lozenge (fig. 101); and as a member having this form we must consequently consider it and regard its relationship with other parts of the flower. Separating the lozenge-shaped member, together with the lines which in part bound the forms to its right and left, we have a wedge and two lines which flow more or less in concert for a time and then diverge. (Fig. 102.) A similar thing occurs

FIG. 101.

FIG. 102.

in that variation of the anthemion seen in Plate X., fig. 6, where we have a wedge-shaped member, in the form of a small conventional coronet, and again have parts proceeding in the same direction for a time, and then deviating to the right and left; and

Fig. 103.

Fig. 104.

in figs. 103, 104, the same principle is applied. The

diamond or lozenge-shaped organ appears to act as a wedge which separates members having a like direction, and gives to the sundered portions new courses.

11. The circumstances which we have here indicated being associated in the mind with fitness and the necessary course of things, seems to contribute much to the apparently spontaneous desire for the alternation of parts.

CHAPTER XI.

ADAPTATION.

1. THE principle of adaptation, or fitness, is of paramount importance in its relation to decorated objects, for the applying to them this quality gives them favour in the eyes of those who use them, and tends to the exaltation of art, while its absence brings ornament into contempt, and thus induces its degradation.

2. In order that art become loveable in the eyes of the people, it is necessary that it in no way militate against the utility of the object which it adorns. If the iron balustrade of a staircase be ornamented with pointed forms which cause the ascending and descending the staircase to be attended with the danger of destroying the dress or injuring the person, or if the enrichment of the door-handle hurt the hand, or the spout of the teapot be so placed, for the sake of ornamental aspect, that the restorative beverage issues from the top of the pot before it runs from the spout, the decorative principle sinks, as the ornamental form brings about a deteriority of the article in relation to utility, and the ornamental aspect is gained at the cost of comfort, safety, and ease.

3. Utility, or adaptation to the purpose intended, must precede enrichment. The most useful form for the object, or most appropriate condition of the surface to be decorated, must be first ascertained, and then the enriching may take place by the application of forms or lines which in no way detract from utility or comfort; but no other form than that which is most desirable must be used in order that a special treatment of the surface be employed, for perfect regard to fitness can alone save art from suffering condemnation.

4. In vegetable nature the utmost regard to fitness is manifested, and those forms only are employed which are in accordance with the requirements of the position; and even the constitution of the plant varies with the surrounding circumstances—hence in relation to adaptation or fitness we must also be taught by plants.

5. The most succulent plants grow in the dryest positions. Our little stonecrop, with its leaves plump with watery juices, grows upon the wall. The mesembryanthemums abound in the sandy deserts of Africa, and the cacti on the steppes of Mexico, and the plants which grow in water contain much air.

6. No one can doubt the adaptation of these plants to the circumstances under which they exist. The stonecrop is located where but a scanty supply of water is obtainable, and where many and long-continued periods of drought are endured; but we yet find this little plant flourishing, as it stores away the water collected during the period of rain in order to meet its constant requirements. This is

also true in the case of the mesembryanthemums and cacti, for in both instances the plants are also called upon to endure droughts which are in some cases even continued for years, and in both instances also the plants have the power of storing up water, the former in its leaves, and the latter in its stems.

7. The extent to which these plants are adapted to the surrounding conditions is further seen when we remember that the vegetable world stands between the mineral and the animal kingdoms—that vegetables arrange the mineral constituents of the earth into organic matter such as is suited to the building up the animal body—that they are indeed the true makers of flesh, which needs only to be arranged by the animal and appropriated to the renewing the wasting body; these facts, together with the knowledge that much water is required in order to the fabricating the animal frame,* show how perfectly the mesembryanthemums and cacti are fitted to administer to the requirements of the animals which dwell amidst them, and are sustained by them. In a land where water is a rare commodity, and is only to be had at particular seasons, and where the land is parched and dry, the higher animals could not exist were it not for the vegetation which furnishes at the same time both meat and drink. The sheep, Livingstone tells us, feed upon the succulent leafage of the mesembryanthemum, and the wild ass quenches its thirst and allays its hunger by feeding upon plants of the cactus tribe; if

* As much as 111 pounds out of the 154 pounds of matter of which the human body is composed, is water only.

the plant selected by this latter animal for a meal be enclosed by spiny armour, which is frequently the case, it, upon reaching the plant, prepares for the attack by separating its lips and taking hold of the long hard spines with its teeth and extracting them one by one, or by knocking them off in a less careful manner with its horny hoof; these plants then are perfectly adapted to the surrounding circumstances and the conditions amidst which they exist.

8. And the adaptation is not only in minor particulars, but arises out of the very structure and habit of the plant which results from the operation of the hidden principle of life. The skin of the succulent plant does not abound in the little openings or pores (stomachs or mouths) which are so numerous in the external covering of other plants—there being as many as ninety thousand of these in the square inch in the case of the under surface of the leaf of the cherry laurel, while the leaves of succulent plants have few openings, if any; hence when water comes in contact with the root, it is sucked in, and is conservated, to be used as circumstances require; while in the case of the plant growing in damper soil evaporation takes place with rapidity—even a sunflower, when full-grown, giving out twenty ounces of water in the day of twelve hours.

9. What has been said is sufficient to show that the plants noticed are adapted to the localities they occupy, and to the purpose they are intended to perform in the economy of nature; and the same perfection of adaptation will be seen if we examine water-plants. The majority of the plants growing in water

are attached by their roots to the bottom of the stream in which they are situated. When this is the case we find the minimum quantity of water embodied in the substance of the plant necessary to its health and well-being; and in the structure of the plant large chambers stored with air, the object of which is apparent. Water is liberally supplied, but should the stream deepen through an excess of rain the plant may be entirely cut off from the air, owing to its attachment with the earth forming the bottom of the stream, hence air is stored up in these instances, it being as necessary to the life of the plant as water.

10. The instances brought forward are sufficient to reveal adaptation to the requirements in the case of habits arising out of the action of the vital force, but there are other manifestations of adaptation, or fitness, arising out of the character of the principle of life, which are manifested in a somewhat different manner.

11. The forms of the parts of the plant are in harmony with the circumstances amidst which the plant exists. The trees which grow highest upon the mountains, and the plants which grow upon the unsheltered plain have long, narrow, and rigid leaves, which, owing to their form, are enabled to bear the fury of the tempest to which they are exposed without injury. This is seen in the case of the fir-trees which grow high upon the mountains, where the leaves are more like needles than leaves as they are commonly presented to us, and also in the species of heath which grow upon exposed moors; in both cases

the plants are, by the form of the leaf, enabled to defy the blast, while those with broad leaves would be shattered and destroyed.

12. But not only is the form of the leaf such as fits these plants to dwell in such inhospitable regions, but other circumstances also tend to this result. The stems are in both cases woody and flexible, so that while they bend to the wind they resist its destroying influence by their strength, elasticity, and woody character.

13. Not only are plants with narrow rigid leaves those which occupy the exposed position, but other plants are also found in such ungenial situations, which have broader and more membraneous leaves. In this case there is, however, a peculiarity manifested worthy of special notice—which is the broad and membraneous leaves of those plants which grow in exposed situations are formed into a rosette at the base of the plant, and are closely pressed to the ground, while those which are situated high upon the stem are in all cases narrow. If the same plant grow in the fertile valley and high on the mountain-side it will, in the one case, have broader and more luxuriant foliage, and in the other, narrowed and skimp leafage, for in the one case the plant is exposed to the fierceness of the boreas, and in the other it is protected from the winds. These instances surely show adaptation to the circumstances.

14. In relation to the stem of the papyrus, Sir William J. Hooker mentions a fact which manifests adaptation to its position. This plant grows in water, and attaches itself to the margins of rivers, streams,

and ponds, by sending forth roots and developing long underground-stems in the alluvium by the sides of the water. Owing to its position it is exposed to the action of the running waters, the influences of which it has to withstand; this it does, not only by having its stems of a triangular form—a shape well adapted for withstanding pressure, but also by having the stem so placed in relation to the direction of the stream that one angle always meets the stream, and thus separates the waters, as does the bow of a modern iron steam ship.

15. In the flower there is the same adaptation to the purpose intended; but before this can be fully set forth it is necessary that we be familiar with the structure of the flower and the parts which enter into its composition, otherwise it will be impossible that we see the extent of the adaptation.

16. The flower, as we have already seen (Chapter X., s. 2), is usually made up of four series, or kinds, of parts (Plate XIII., fig. 2)—an outer ring of leaves, which is most commonly green, and consists of a varying number of constituent leaves, from two to several, which may either be separate or joined; an inner ring of leaves, which is usually of a bright colour, and composed of the same number of members as the outer ring of green leaves, which may also be united or separate; a ring of slender, threadlike, or awl-shaped members which are terminated in yellow knobs, and a series of green parts which occupy a central position in the flower; in the case of the third series the members are most frequently separate, but they may be united by their

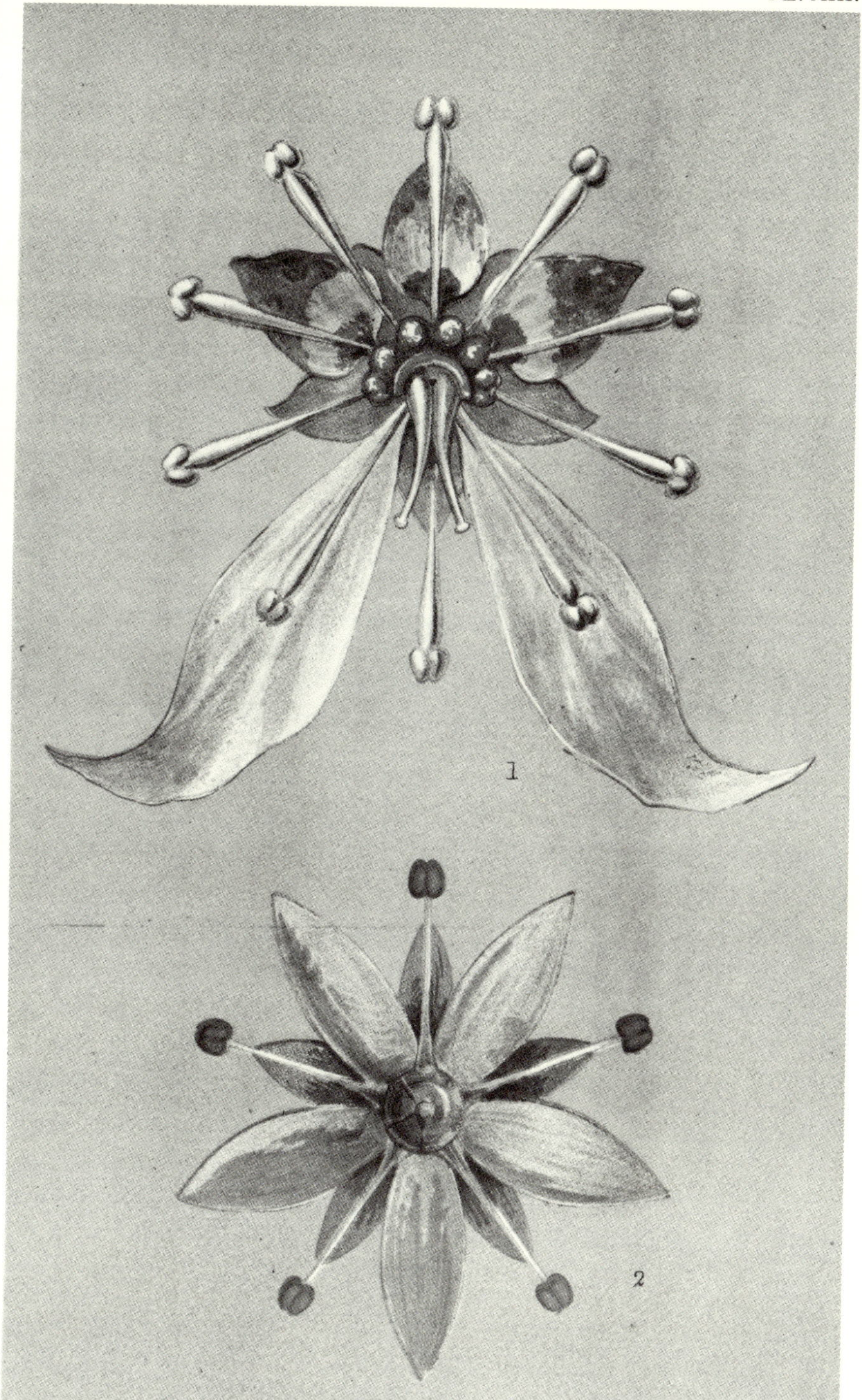
1
2

stalks (fig. 105), or by their knobs (fig. 106), while in the case of the central bodies union frequently takes

Fig. 105.

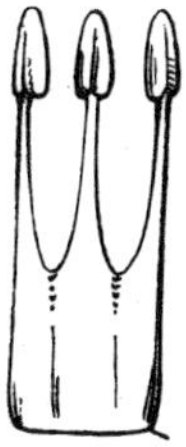

Fig. 106.

place so perfectly that the central series seems to be but one member, while it is yet formed of many portions. In respect to the object of these parts in relation to the economy of the flower, it should be noticed that the two external series—the leafy rings—are intended to envelop or protect the inner portions, which are more essential to the well-being of the plant, and that the inner portions are destined to bring about the production of seed. Within the knobs which terminate the members of the third series (figs. 74, 91, and 105) of parts, a quantity of pollen, or dust, is formed, which, when examined by the microscope, is seen to consist of little living spherules or cells, which are the most simple form of plant; this dust is scattered by the bursting of these knobs, or chambers in which it has been generated, when a portion comes in contact with the central members of the flower, and here, finding genial food formed of honey, gum, and other mucilaginous forms of matter, sends down roots in the midst of the substance of these central portions, by which roots seeds are generated in the manner that potatoes are produced by the underground stems of the potato plant, only in the case of the potato several are formed upon one stem, while in the case of the seed but one is formed, and that on the end of each suspensor. This is a popular setting forth of the floral parts and of the works which they perform in the economy of the flower, and is sufficiently accurate and detailed for our purpose, but it should also be noticed that if water comes in contact with the fine dust (pollen) of the flower its fertility is destroyed, as contact with water causes it to burst, and thus brings about its destruction.

17. To return to our subject, we find that flowers are usually inverted, as the harebell (fig. 107), and many species of campanula; or they only open when there is every prospect of fair weather, as the majority of our garden flowers; or the flower closes at the approach of rain, as the shepherd's weather-glass, or pimpernel (*Anagallis arvenia*), and minor convolvulus; or the head of flowers closes as the sky becomes overcast, as the daisy (*Bellis perennis*) and yellow hawkweed* (*Tolpis barbata*); or if the flower opens, the little knobs do not shed their pollen or dust till the day is bright, as in the apple and pear; and if the flower grows in an open and exposed position, as upon a mountain or moor, it is frequently elevated upon a flexible slender stem, which, by yielding to the wind, enables it to keep its back to the storm, and thus that protection which is needed is afforded.

18. To take one or two individual instances, the little

* What is usually termed *the* flower in both these instances is a head of flowers.

harebell, with its nodding cup situated at the upper extremity of a long and slender elastic stem, studs the moor and rises amidst the bushes to a height greater than usual; here we find the heart-shaped leaves, which are fast yielding to decay, disposed in the manner of a rosette pressed upon the ground; then we have the stem, with its narrow elongated leaves; and, above all, the flower in the form of a nodding bell (fig. 107). How appropriate is this struc-

FIG. 107.

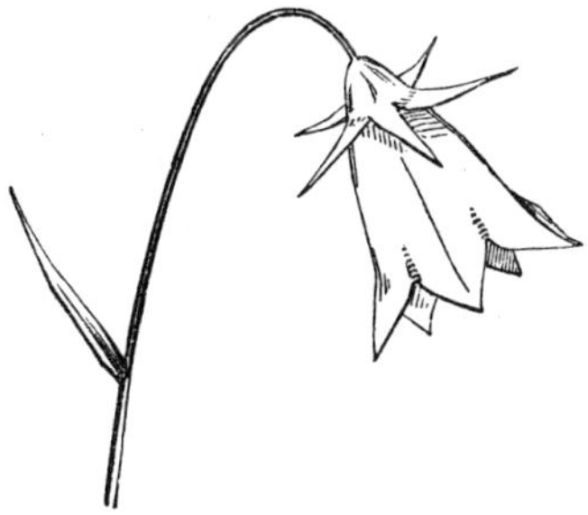

ture to the circumstances in which the plant is placed; the drooping bell shelters the fine dust from the rain; and when the flower is exposed to the beating water-drops and is yet open, the pollen is protected through the back of the flower being kept to the storm owing to

FIG. 108.

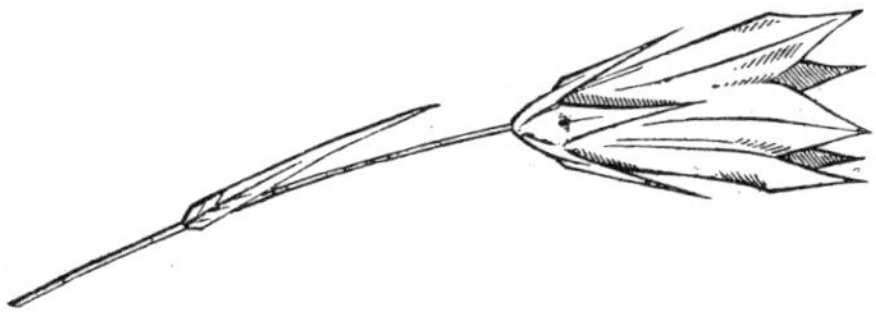

the slender and flexible character of the stem (fig. 108); here, then, is an instance of special adaptation.

19. A yet more striking example of adaptation occurs in the case of the common edible pea blossom, where we have the five little green leaves, which form the external series of floral parts, united into a cup-shaped socket, which invests the base of the flower (fig.

Fig. 109.

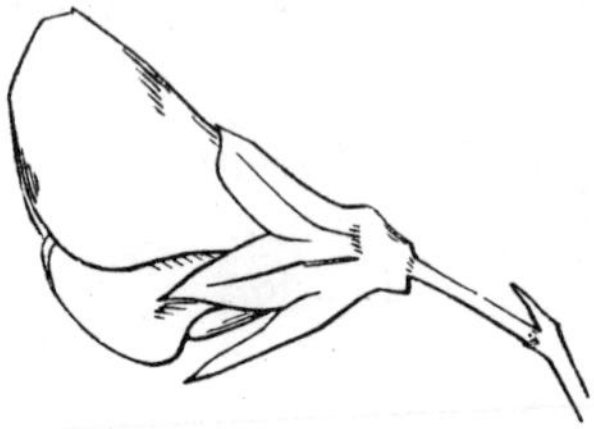

109); within this we have five white leaves, which are of variable forms and sizes; the upper is large and is folded in the centre (fig. 110); the lateral, or side

Fig. 110.

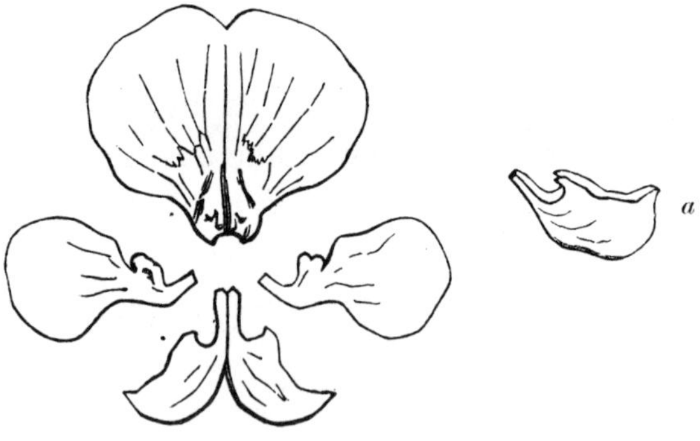

leaves, which correspond with each other, are smaller, and are situated within the margins of the upper sheathing leaf, while the two lower are smaller still, and unite to form a little scoop-shaped member (fig. 110, *a*). Within this lower scoop-shaped portion the ten organs, which unitedly form the third series of floral parts, are situated, and of these nine are united into one bundle by a consolidation of their stalks throughout a great portion of their length, while one stands alone (fig. 111); and within the little scoop formed by

FIG. 111.

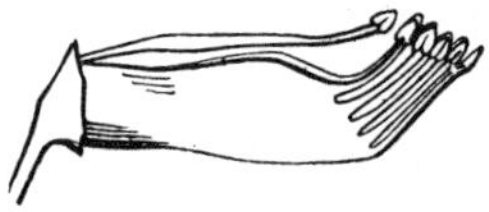

the nine united stalks the central organ rests, which in this instance consists of but one portion: we here have the structure of the flower; the inner or central member, we see, rests in a scoop-shaped organ formed by the union of nine stalks, which are each terminated by knobs, and upon the upper portion of this central organ the solitary unit of this third series rests, which is also crowned by a golden ball. This mass, consisting of the central organ and the ten members of the next series, is concealed in a second scoop-like organ formed by the union of the two lower white flower-leaves, and these are protected by two overlapping white leaves and a large upper leaf, also white, which sits on the others in the manner that the ridge-tile sits on the summit of the roof of a

FIG. 112.

house (fig. 112), and the base of the flower is further protected by the green cup at the back; we see, then, the special provision made for the protection of the pollen or fine dust of this flower, and thus for its fruitfulness or seed-bearing.

20. Not only is protection afforded by part overlapping part, but the whole flower is turned by the wind in the manner of a weathercock—hence it also keeps its back to the storm.

21. We now see something of the extent of the protection afforded to the pollen of this plant, and this careful concealment is ever vouchsafed unless the weather is very propitious, when the large upper leaf is uplifted, and the side leaves are spread, and the lower scoop-shaped member is depressed. But even in this spreading of the parts there is a special object: the bee can now enter the flower; and in its search after honey it brushes past the ruptured knobs, and thus conveys the pollen to the central organ of the flower, where its growth will be fostered and its development ensured; but should there be the promise of a shower, the little basal scoop is uplifted, the lateral leaves doubled together, and the large upper leaf is lowered and folded over the other parts, and unless the pollen has fulfilled the purpose of its creation, should the rain beat and the wind blow, the back of the flower is constantly directed to the storm, and thus the parts are preserved from injury.

22. Two circumstances, which we have not yet mentioned, combine to give to this structure of the pea-blossom exceeding interest; first, the seeds of the different kinds of pea contain more nourishment than any other food which we use; and second, the large upper sheathing-leaf, which is primarily instrumental

in furnishing the perfect protection of which we have been speaking, can be shown to occupy an abnormal position, its place being that indicated by the dotted line in our diagram (fig. 112); these facts, considered in conjunction with the structure of the flower already noticed, manifest most clearly the recognition of the principle of adaptation or fitness, in the case of plants, to the work which they are intended to perform and the circumstances in which they are called to exist; and as in plants so in ornament, adaptation and fitness cannot be too strongly manifested or too fully considered.

23. We see that the study of the vegetable kingdom suggests the utmost regard to fitness and perfect adaptation; but in order to apply the principle to the requirements of ornament we shall first contemplate one or two well-considered objects furnished by ancient art.

24. Professor Semper used to say, that the history of a nation can be read in the form of the water-vessel. Without insisting upon the accuracy of this assertion when taken in its wider significance, we yet say that a great deal may be learned from such a simple utensil; but in order to this, the water-vessel must be a well-considered object, and be perfectly adapted to the purpose for which it has been originated.

25. The characteristic form of the Egyptian water-vessel is represented in fig. 113, and of the Greek in fig. 114, and while both answered a similar purpose they yet widely differ in shape.

26. But not only do these vessels differ in form, but associated circumstances differ also; and it is this

variation of circumstances which brought about the difference in form of the two water-vessels. The

Fig. 113.

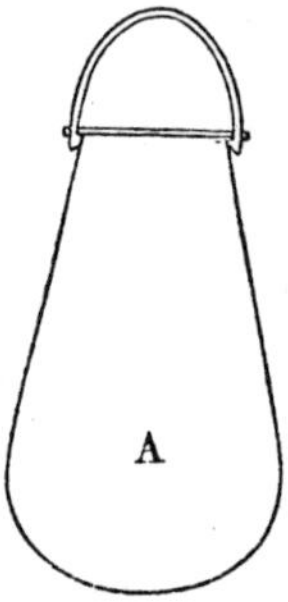

Fig. 114.

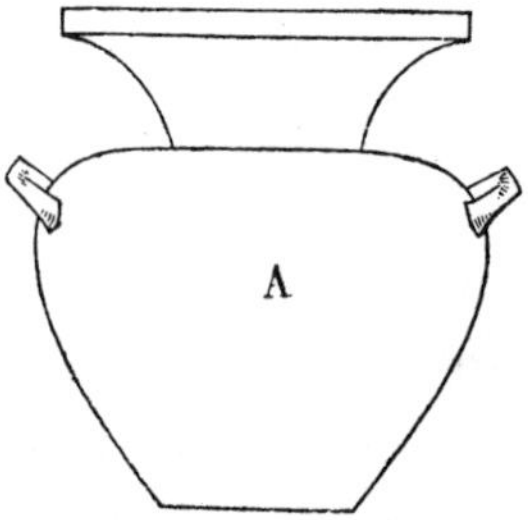

peculiar features of the Egyptian water-vessel are its formation out of bronze, the roundness of its base, which renders it unfit for standing, the narrowness of its mouth, and the handle arching the orifice; and of the Greek its being wrought of clay, the secure base, the wide mouth, the contraction in the centre, and the handle at either side. We learn from these forms of vessel, allowing their adaptation to the conditions which they are intended to fulfil, that the Egyptians drew water from a river, or at least from a position which required the vessel to be attached to a cord and cast into the supply, for the roundness of the base at once points to this, it being a provision for enabling the vessel to fill by turning upon its side rather than float upon the surface of the waters, which it would readily do were its base flattened, and its formation out of metal is to facilitate this end. The arched handle not only points to the attachment of the vessel to a string in order that it be cast into the water, but also to the carrying the vessel pendant from the hand, in the manner that the pails of the present day are carried, and the contracted mouth restrains the splashing over of the water; and what this simple water-vessel points to we find to have been the case, for the Egyptians derived water from the Nile in the very manner that we have shown the vessel is adapted to operate. But with the Greeks habits are different, as the water-vessel manifests. The base is flat, in order that the vessel may stand; the mouth is large in order to collect the water which falls from above—from the dripping rocks and water spouts—and this largeness of orifice points unmistakeably to the collecting falling water. This being the case, a vessel formed of heavy metal was unnecessary; the contraction prevents the water splashing over when carried, and up to this point the vessel was filled and not higher; and the handles at the side show that it was carried on the head. But in conjunction with this mode of carrying there is another consideration of interest, which is, the centre of gravity must be high; if we attempt to balance a stick with one enlarged end on the finger, it will be found necessary that the weight be at the top, and in order that the object ride steadily and easily, it will be necessary that the weight be considerably elevated. In the Greek water-vessel, which is to be balanced on the head, we find this condition fully complied with, the centre of gravity occupying the position indicated by the letter A in the illustration, while in the Egyptian vessel the centre of gravity is low, as is also indicated by A; but where the vessel is to be carried underhand it is as great an advantage to have the centre of gravity low, as it is in the case of a coach, where insecurity results from its being raised, and that in direct ratio to its altitude. The Greek water-vessel, then, consists of a capacity or chief cavity, a funnel to collect and guide the water, a base for the vessel to rest upon, and handles to enable it to be raised to the head, and the centre of gravity is high in order that it may readily balance—and we read from this vessel that the Greeks procured water from dripping rocks, water-courses, &c., and this is exactly what did occur, for the people derived water from these sources in the manner set forth. These are the direct teachings of the Egyptian and Greek water-vessels; but how many circumstances and incidents of common life can be conceived to grow out of these modes of collecting water. There is the gossip round the well and the lingering by the river side, where the image of the date palm is mirrored forth by the glassy surface of the waters. The effect of the noise of the splashing waters, which seem to be filled with the energy of life, upon the mind in the one case, combined with the loud and energetic speaking which would be necessary in order that the voice be not drowned by the noise, and of the calm tranquillity of the river bank in the other, where the limpid stream ever flowing on in silent majesty, must be considered. Then we have reference to the potter's art in the one case and the metal-worker's in the other. But we will not push the subject further. We bring forward these two vessels in order to illustrate the manner in which well-considered objects speak of the habits and customs of the times and nations in which they originated.

27. We have illustrated the extent to which ancient

objects were adapted to the purpose for which they were intended. Let us now see the application of the principle of adaptation to modern requirements.

28. An object should not only be so far fitted for the work for which it is intended, that it is possible to use it for the purpose of its production, but it should be perfectly adapted to meet all the requirements of the work to which it is assigned, and that in the easiest and most simple manner.

29. The principle of fitness is constantly violated in the placing handles and spouts on vessels; but this arises out of the want of a knowledge of those principles involved in the attaching such affixes to utensils. What can be worse than having the handle so placed in relation to the spout—say in the case of a teapot—as to give a considerable leverage, which can only be exerted against the person who is pouring out the tea? We constantly hear ladies complain of the exceeding weight of the teapot, and of the fatiguing and wrist-breaking work of pouring out tea, while others have teapots of equal capacity which are not heavy, and which do not cause this operation to be either wrist-breaking or fatiguing; yet such teapots are perhaps the exception in these days. A pound weight may balance a half-hundred weight if the leverage be in its favour; so a light teapot and a small quantity of liquid may practically have a considerable weight, when the teapot is held by the handle, if the leverage is exerted to the disadvantage of the person who sustains it. Can any one, however, say that the object is perfectly adapted to its work if in this or any other particular its use is attended with difficulty

or disadvantage. If the handle is a disadvantage we had better be without it, and hold the vessel by merely sustaining it between the hands, protected from the heat by two kettle-holders.

30. In order that the handle and spout be rightly placed on a tea-pot or similar vessel, it will be necessary that the form be originated with special reference to these affixes, and with full knowledge of the principles on which these additions are to be made. The first thing to be ascertained, in order that the spout and handle be rightly situated in relation to each other, is the position of the centre of gravity of the general mass of the vessel (. *a*, fig. 115); this being found,

Fig. 115.

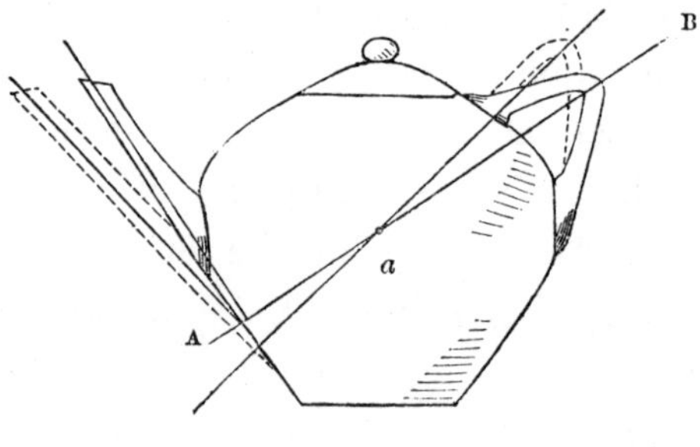

the centre of the handle falls upon any right line (A B, fig. 115), which passes through the centre of gravity (. *a*), but that which is of special importance is this —the spout must also fall upon a line so situated that it forms a right angle with the line passing through the centre of gravity and the centre of the handle, as the line A B. If the handle is raised to the position indicated by the dotted line the spout must be lowered to the position of the dotted spout, for the direction lines of the spout and handle must bear the relation to each other of the lines enclosing the right angle. If the angle be made obtuse, leverage tending to bring about the lowering of the teapot will be induced, and if the angle be made acute, pressure will have to be exerted in order to depress the mass; hence it is of the utmost importance that the relation of right angle be maintained between the spout and the handle.

31. In some instances we find the tea running from the spout of the teapot before the pot itself is full, this of course arises from having the spout short; but a more common evil is the tea runs over the top of the pot before it runs out of the spout, when an attempt is made to pour out the liquid. In order the more fully to prevent this, we see no objection whatever to the lid having the form and situation shown in fig. 116, if necessary, for fitness and adap-

FIG. 116.

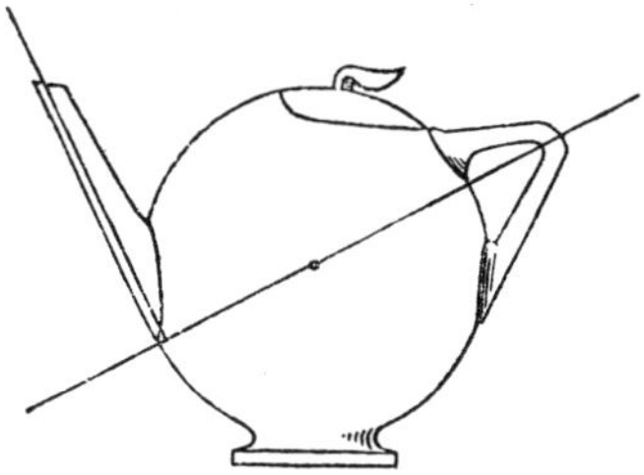

tation must be a first consideration in the production of useful objects.

32. In respect to the placing handles on jugs nothing can be worse than our common practice of attaching high handles to vessels of this class, which are without the advantage of a spout deviating in direction from the vessel. When we take hold of the water-

ewer in the bedchamber we grasp the handle near its base, just where the handle ought to be situated; and were it not that the handle is continued to the

Fig. 117.

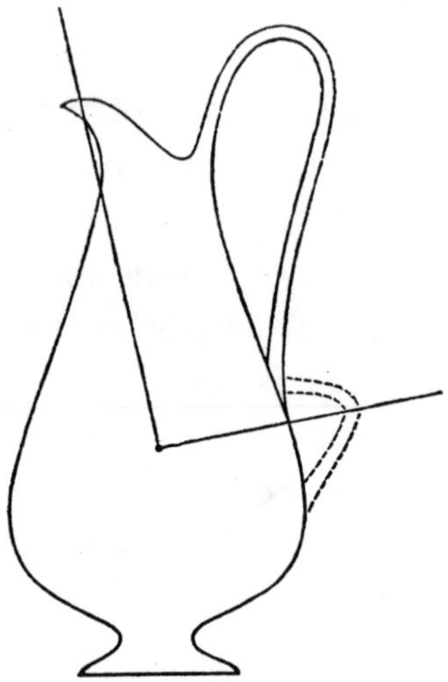

lower portion of the jug it would be as difficult to pour water from it as it is to raise a chair by taking hold of one of its legs. Fig. 117 represents a

Fig. 118.

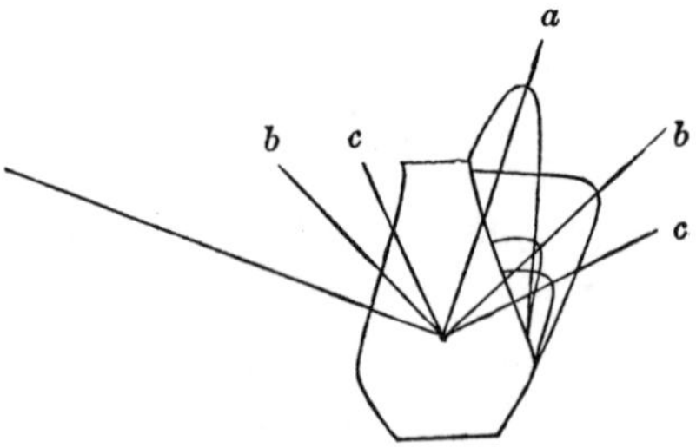

jug, and the dotted handle marks a correct position for this affix. Fig. 118 also gives positions for the handle and relative positions for the spout; thus the handle *a* and the spout *a* correspond; the handle *b* and the spout *b*, and the handle *c* and the spout *c*.

33. A consideration of the principle of adaptation will further show us that each material is especially susceptible of some particular treatment, and this will reveal to us the folly of attempting the imitation of one material with another. Clay is especially adapted for being formed into jugs and other vessels of capacity which are required to hold water; why, then, cause clay jugs to imitate wicker work, as though *it* were to any extent adapted for fabrication into vessels which of necessity must be water tight? The folly of supping soup with a knitting-needle, of lading water with a gridiron, or carrying wine in a sieve is not greater than that of causing a jug to appear as wicker-work. Neither is the making a jug in imitation of cooperage much less a folly; still, a barrel will hold water, but why make an earthen vessel look like a specimen of bad cooperage and of defective wood carving when the clay can be wrought into objects of capacity of infinitely greater beauty than those formed of wood? Wood has its special uses, and is peculiarly suited to certain works, and clay is adapted for the formation of special articles; then one material should not stoop to do badly what the other does well when it can do well what the other cannot; the effort must be to find the susceptibilities of each material, and to apply the utmost skill to the furtherance of its legitimate uses.

34. Not only must the susceptibilities of the material be ascertained, but the most appropriate mode of fabricating the material into forms of utility and

beauty must be sought out. Clay is peculiarly fitted for being worked by the potter's wheel, as this mode of fabricating it into forms of utility is simple, and is at the same time calculated to bring about the production of graceful forms, for here the influence of the attraction of the earth upon the semi-plastic clay aids the potter in producing favourable results, as will appear from the following—A vase may be said to consist primarily of a capacity and a double funnel (an inlet and outlet). The former we represent by an ellipse, the latter by a cylinder (fig. 119). We now determine the width

Fig. 119.

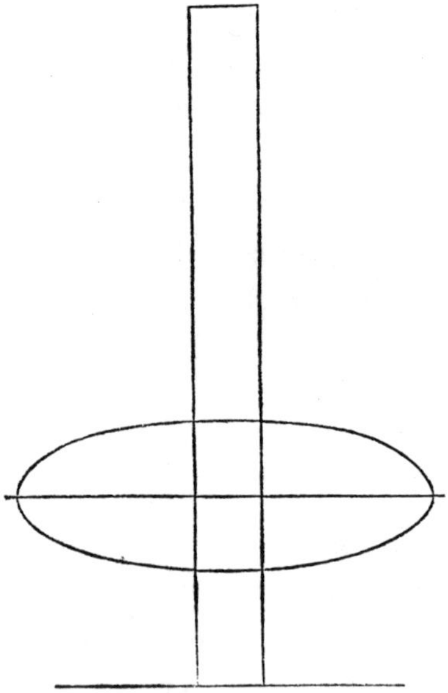

of the foot at varied heights, at *a* and *b*, fig. 120, in this case, and then connect the parts by the plastic clay, which, by yielding to the earth's drawing force, takes the direction given by the dotted lines. The

FIG. 120.

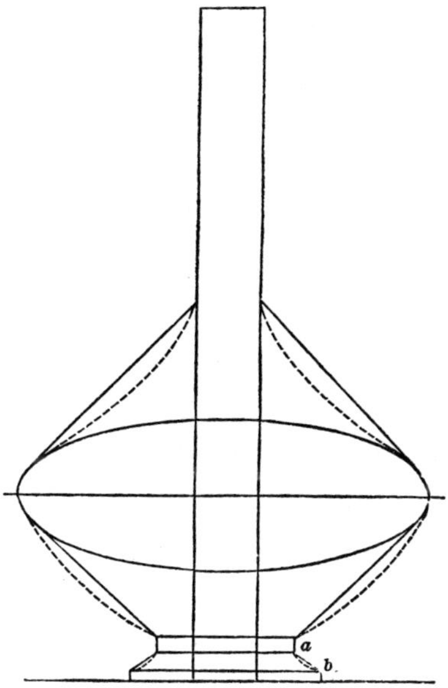

FIG. 121.

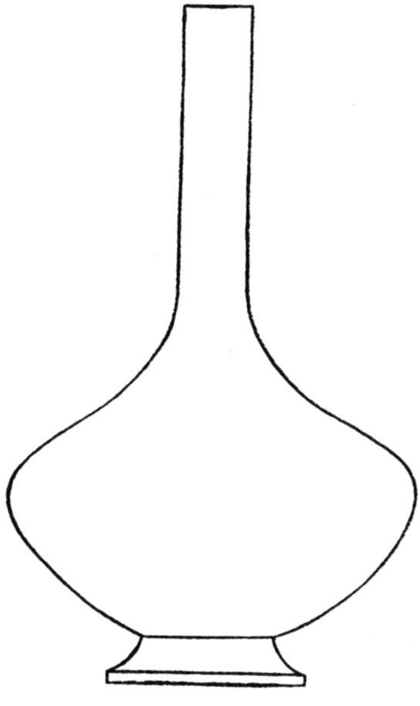

form thus produced separated from constructive forms is seen in fig 121.

35. Another illustration of the favourable influence of attraction upon plastic or semi-plastic matter in bringing about the production of beautiful forms occurs in the case of the flask where pressure of air from within causes the distension; but the form results from the attractive force of the earth acting upon the plastic glass and the included dense air, which is also wrought upon by the same drawing influence. Some of the smaller vessels of the Greeks manifest forms which would appear to result from vessels of certain shapes being filled with fluid while yet plastic, which also shows the manner in which attraction aids in the production of vessels of utility from clay; and this one fact — that clay is susceptible of this attractive influence and is placed under its direct operations when wrought into form by the wheel,—tends to show that this is the most appropriate mode of working it, and that it, as a material, is especially suited to this method of fabrication.

36. We see that each material can be worked into a form which is to it especially befitting, and that vessels appropriated to special purposes have a limited range of form within which they may vary, and yet be perfectly adapted to the work which they are intended to perform, but departure from a typical shape beyond certain limits is attended with the sacrifice of utility. The Doric column (Plate I.) has a form which is universally admired by those who have cultivated their sense and perception of the beau-

tiful, but it arrived at its just proportions after the lapse of many years. Age after age tried only to perfect its form, and not to originate that which was new; so we must strive to discover the best class of form for the object, and to perfect an individual shape till its beauty is impressive. So long as we are striving after new shapes and are forced by this excessive desire for novelty to produce jugs in the form of the human head, we shall make but little progress. Allowing that men's heads without brains are common, we need not imitate such, especially as jugs—for no one will argue that such is a befitting form for a vessel of this character.

37. Another class of folly often manifested, arising out of a want of adaptation, is the making one surface or fabric imitate another surface or fabric—as silks to imitate the coarse texture of the Turkey carpet. In the case of this coarse material, the best curves are formed that the material admits of; hence, in silks, we should strive to form the most perfect lines and shapes of which the material is capable, and not to imitate the defects of a coarse material. The effort should always be to reach the ideal perfection, and not to perpetuate a defect or deformity. Then we make wall-papers imitate woven fabrics, and carpets represent marble floors; while, if our floors were marble, we should cover them with rugs in order to increase our comfort. All these, and a thousand other absurdities, arise out of the want of a due consideration of the principle of adaptation or fitness.

38. A consideration of that adaptation which prevails in the more perfect works of ornamental art

of all ages, and of that fitness which occurs so fully in the diversified vegetable objects, leads to a number of other thoughts, all of some interest; but upon none of these can we dwell, and but few can we name, yet some must not be passed without notice.

39. In the case of a plate, the border only should be decorated, for this is an article of utility, and the centre is covered when the plate is in use; and the interior of a cup should never be enriched with decorative forms, save just round the orifice, for to do more is to be extravagant. With wood carving, as the legs of tables, the legs and backs of chairs, &c., no small points must jut out, as there would be the constant danger of their being broken off; and, in the case of metal-work, no sharp prominences must occur which are calculated to injure the person or destroy the garments of those who make use of the article.

40. Walls should at all times be flat; and the very appearance of rotundity should be avoided, save in those parts with which the body does not come in contact. The cornice round the upper portion of the room may most desirably be in relief; but this is purely an architectural feature; and even the wall, down to a certain level, may be enriched by a relief diaper, as occurs in the Moorish buildings, and is rendered possible in the case of common walls by an improvement effected by Messrs. Scott, Cuthbertson, and Co., of Chelsea, London, who furnish "self-colour" papers, in low relief, of a very beautiful character. But in all cases flatness is a pleasing feature where the body comes in contact with the surface; hence this character should be rigidly preserved, for it is the prerogative of art

to take hold of the most pleasing feature of an object, or most welcome condition of a surface, and to impress it, adding with it that amount of enrichment which is desirable. It may be thought that in the case of the Moorish treatment the belt of smooth tiles (corresponding with our old-fashioned wainscoting) is not sufficiently high to protect the body of the occupant of the apartment from contact with the embossed portion of the wall; but a moment's meditation shows that it is—for it rises above the elbow when the man is standing, and above the head when sitting. For, where Moorish architecture occurs, the low settee and ottoman are used, which raises the occupant but little above the floor of the room, and a reclining position of the body is adopted.

41. In the case of carpets and other floor-decorations the objection to relief, or the appearance of relief, becomes greater than in the case of the wall, for nothing can be more uncomfortable to walk over than a rough and uneven surface of land and water, bestrewed with bushes and other large objects which cast massive shadows, interspersed with fragments of Louis-Quinze ornament. Certainly, in the case of a floor at least, flatness is a desirable feature, and if desirable, then a feature which art should impress. A new invention in abominations has recently occurred; a carpet having appeared which is enriched with well-formed bows of ribbon, in which the light and shade are so well managed that the aspect of relief is fully given; but the peculiarity and crowning point rests in the centre of each bow being securely attached to the ground by a large button, hence the

uncomfortable character of the pattern is increased to a maximum by the idea being afforded that those who walk over the carpet are proceeding through large high loops which are securely attached to the floor.

42. In setting forth the principles we now advance we are opposing a tide of feeling which has its chief advocate and expounder in Mr. Ruskin—yet a tide which, happily for decorative art, is waning. This gentleman may dogmatically say that no one who can paint a picture would decorate a wall with ornament.* But to apply this summary method of dealing with the case to the opposite side, may we not with equal truth say that no one who could justly decorate a wall would paint a picture upon it? But it is easy to dogmatize and make bold assertions, it is another matter, however, to offer proof and to build up statements by reason. As a decorative scheme, the Alhambra is in our judgment much higher than the Vatican, for in the latter the *ornament* is bad and the general effect seems sacrificed to the exceeding power of the pictorial works; but while the representative of the Natural School seems to have a true feeling for pictorial art, he has little knowledge of ornament, and while he has done much to advance the principles and maintain the claims of the former, he has done but little, we fear, to foster a genial growth of the latter; for if we take away the elegance from his pen, and his noble power of utterance, we have little practical information afforded respecting ornament, and little advanced which is likely to prove of permanent worth.

43. To talk of walking over beds of roses is very

* *The Two Paths.*

agreeable, and poetical also; and by expressions of this character it is easy to make a successful appeal to tender hearts and feelings, but after all the judgment is not reached. The fancy of a brute may be tickled, but man is endowed with reason which requires to be satisfied. But the case is altered by the fact that not one-tenth part of the designs put forth upon the principles of the Natural School are rosy beds; many are composed of flowers interspersed with wooden-looking lumps of Louis-Quatorze and Louis-Quinze ornament; many more are made up of trees, bushes, and ponds, while others are of the ribbon character. Then we have groined-roof patterns, Gothic-moulding patterns, and many other varieties of a somewhat similar class. The strongest advocates of relief floral patterns need not contend in favour of the walking over inverted groined roofs, but if the principle of light and shade be permitted in one case it must be allowed in others; yet we are ready to admit that the want of pleasing associations may render the latter class of pattern more inappropriate than the former. The only two complete representations of nature in the kinds of pattern named occur in the strictly floral pattern and the pictorial design which embraces in its parts trees, avenues, and lakes. Now it is argued that as what is here represented is natural, and as lakes and trees appear in creation with light and shade and diversified tints, it is legitimate and right to depict them on our floors. This argument seems to have many applications. On the same principle we might say that because a thorn is bounded by curves of exceeding beauty and has light and shade, that it is an agreeable thing to sit upon. It does not follow that because a thing is beautiful where it is that it is a just decoration if removed to a new position amidst altered circumstances. The expounder of the principles of the Natural School speaks much about the "interpreting nature." This is, unquestionably, right; but it does not hence follow that we are to accept his interpretation, for to us it seems that such a rendering is in some cases at least not very accurate. In the case before us we think nature does not teach the application of landscapes or any incongruities to the purposes of floor-decoration, but sets before us a thousand instances in which she most palpably manifests a perfect adaptation to the conditions under which she exists; and when she forms a bed such as we love to walk upon, she fabricates it of the short and tender grass blades which constitute the velvet pile of nature, and has its analogue in the mossy surface of our carpet. In wandering over a common, we instinctively avoid furze-bushes, pools of water, stumps of trees, and the like, hence why place them on the floor where we are compelled to walk? Whenever we come in contact with an object on the moor which casts a massive shadow, we avoid it; then why place it in the path which we must tread? Those who insist upon the application of flowers in relief to the purposes of ornamentation seem only to have looked at the surface of things, and have not contemplated the deep teachings of vegetable nature as to adaptation. In the case of the leader of the Natural School, the knowledge of nature seems to have been derived from a secondary source, for, if we are rightly informed, application was made at South Kensington for the titles of the botanical writings of the author of these pages, of which good use seems to have been made, judging from certain principles and illustrations set forth in the first chapter of the last volume of *Modern Painters;* but in order to the receiving the full teachings of nature reference must not only be made to works written with one distinctive object, but to the volume of nature which is alike open to all. The best lessons which nature teaches are in some cases hard to learn and difficult to acquire, but he who would know her fully must be content to encounter difficulties and spend long hours in ascertaining small truths, and he who does so will find that plants teach consistency.

44. There is another phase of the subject. Plants not only manifest certain forms and combinations of lines and colours, but they set forth a principle of life and energy which in the tropical flora is extremely manifest.* We have now to consider whether this principle of life can be set forth in conventional forms, and if it can, whether a composition in which the spirit of nature is embodied does not more fully represent the plant—or, at least, the principles of plant-growth—than the ordinary floral patterns? and to decide this matter we can call in a high authority.

* The magnificent set of stereoscopic views of Java recently published by Messrs. Negretti and Zambra, set this forth with exceeding power. The energy displayed by the publishers of these views in the procuring photographs of scenes and lands of the deepest interest, and the great cost at which these are procured, should elicit for them that patronage which the undertaking deserves. Figs. 4, 65, 86, are taken from the Java series alluded to.

Some few years back the author, while in conversation with Dr. Lindley, who is well known as a botanist, called attention to the sectional forms of certain seed-vessels as ornamental objects, when the learned Doctor related an anecdote which had occurred in his history shortly prior to this time. He required a wall-paper, and went in search of such to a large shop in Oxford Street, when a number of floral patterns of the highest order were unfolded, but all failed to charm; impatience now began to manifest itself on the part of the Doctor, who said, "If you have nothing which contains more nature than these I must go elsewhere," whereupon a number of conventional patterns were introduced, and one was speedily chosen: here we have a man who has spent his life in contemplating nature, bearing testimony to the fact that by certain conventional forms more nature is manifested than by any of the so-called natural patterns. There is a hidden principle of life in plants unfelt by the casual observer, which lies far back beneath the forms of her parts, as the soul dwells in the recesses of man's nature, and this it is which few can portray.

45. We have set forth our principles concerning adaptation in ornament, principles which are alike manifested by nature and by art in her best periods, hence we have arrived at a stage at which it will be well to advance the principles of a few men whose names are honourably known in connexion with ornament, in order to show how the matter stands; for while we are perfectly aware that no cause is proved to be right by the multitude being on its

side, it is yet well that we make ourselves acquainted with the weight, in the form of great masters, on each side. On the one side we have Mr. Ruskin, who was formerly upheld by Mr. Redgrave, R.A., then a leader of the Natural School, but a man who has since become the expounder of orthodox views and the awarder of merit upon the same sound principles,* and is now in a remarkably unsettled state of mind, judging from the works which have recently issued from the South Kensington establishment, at which he is the Art superintendent. Besides Mr. Ruskin we have no great name on this side with which we are acquainted, and certainly not the name of any great ornamentist. On the other side, we have Owen Jones, a man who, by his works, has manifested his knowledge of ornament, and who has given forth more real practical information respecting the decorative art in his Propositions, formerly sold by the Department of Practical Art for one penny, and now united with his *Grammar of Ornament*, than Mr. Ruskin has in his much more voluminous writings. Hear what Owen Jones says (these passages are two of the Propositions):—"Flowers or other natural objects should not be used as ornament, but conventional representations founded upon them sufficiently suggestive to convey the intended image to the mind without destroying the unity of the object they are employed to decorate." "As architecture, so all works of the decorative arts, *should possess fitness*, proportion,

* See Redgrave's *Report on Design*, bound with the Jurors' Reports on the 1851 Exhibition, and published separately by Chapman and Hall, a most valuable essay.

harmony, the result of all which is repose." Then we have Digby Wyatt:—"Infinite variety and unerring fitness govern all forms in nature." Vitruvius:—"The perfection of all works depends on their fitness to answer the end proposed, and on principles resulting from a consideration of nature herself." Sir Charles L. Eastlake:—"In every case in nature where fitness or utility can be traced the characteristic quality or *relative* beauty is found to be identified with that of fitness." A. W. Pugin:—"How many objects of ordinary use are rendered monstrous and ridiculous simply because the artist, instead of seeking the most convenient form, and then decorating it, has embodied some extravagance to conceal the real purpose for which the article has been made." And the Kensington Department of Art avowedly; but how this tallies with their late practice we cannot see, for the last mania has here been for modern Italian relief ornament, as suggested by the Majolica ware. This enumeration of authorities will give some idea of the weight on either side.

46. Before dismissing this subject, we may have reference to remarks frequently made condemnatory of manufacturers who produce articles in bad taste, for it is said, if such articles were not produced we should not buy them. This is not fair; the producer and the consumer must share the condemnation: so long as the consumer, by his want of knowledge, creates a demand for articles of bad taste, that demand will be met by the producer, for the manufacturer and consumer are alike ignorant of the principles of true ornament,

and the producer concerns himself only with the supplying that for which there is a demand. If it be urged, on the one hand, that in the case of the manufacturer's ignorance of art he should employ some one of taste to see to these matters, it must be admitted that it is equally imperative on the consumer who has no art knowledge to employ some person of sound judgment to buy for him, but how rarely is this done! We have known but one instance in which a person has employed another to buy for him, owing to his want of taste. Owen Jones seems to be right when he says, "No improvement can take place in the art of the present generation until all classes, artists, manufacturers, and the public, are better educated in art." A curious fact occurs in relation to the partners of one of the largest carpet manufactories in the north of England and their rugs, which is appropriate here. The author of these pages, being on terms of some intimacy with one of the partners, upon calling at the works found hung upon the walls of the counting-house, as pictures, certain of their magnificently-wrought lion and similar rugs; here they were appropriate and effective. But upon visiting the house of this liberal and princely manufacturer, in order to partake of his bounties, it was found to be a Gothic mansion, well stored with the most appropriate Gothic furniture, and nowhere was a lion rug to be found. Here we have an instance which shows that so long as there is a demand the supply will be kept up, even while the manufacturer holds contrary views of art; and indeed it is better that the people, through their knowledge, refuse to buy works of false art than that they be debarred from procuring such simply through the cutting off the supply. The former state of things we may hope to see; the latter we must never hope for.

47. Not only is a flat treatment the only treatment which is adapted to a flat surface, as a wall, but there are special effects which are peculiarly suited to the particular rooms; thus a drawing-room should be bright and light, and calculated to cheer and encourage pleasurable conversation—hence no dark paper is suited to such a room, for it would fail to induce the condition of mind desired. The paper on the wall of the dining-room should be darker and lower in tone, for the object of special attraction here is the spread table—hence no object should call attention from it. The library should be rich and calculated to foster meditation, and the bed-room cheerful and calm, so that in the hour of sickness the influence may be soothing. Yet in all cases both the wall and the floor, hence the paper and the carpet, should serve as a background to the contents of the room, and should be of that subdued character which enhances the value of the furniture exposed. If good engravings or sketches are hung upon a wall covered with an unsuitable paper, they will hardly ever be noticed; but if they be hung upon a suitable background their attractiveness will call forth many a comment, hence we see that a special class of paper is peculiarly suited to a particular condition.

48. A consideration of adaptation or fitness further shows that ornaments destined for vertical and hori-

zontal positions—for walls and floors—should be of a different class; that the wall decoration should have its halves corresponding while the floor should consist of radiating parts.

49. This is a principle which has been universally obeyed in the best periods of art, and occurs most fully in the vegetable kingdom, as will be seen by attentive observation. The top view of every plant (with the exception of a few so low in the scale of organization that they are not popularly regarded as plants) shows it to be of a radiating character. While this statement is true, even in the case of trees and the larger vegetables, it concerns us chiefly in relation to the smaller plants. In wandering over the moor we notice that every plant which we look down upon has this radiate structure—the daisy, the buttercup, the jelly-flower (*Cardamine pratensis*), the primrose, the cinquefoil (*Potentilla*), and whatever plant we look down upon, it matters not what it is, for in all cases the leaves and the flowers and the prostrate stems spring from a common centre. These are the horizontal ornaments of nature, and with the top or radiate view of these structures are we alone familiar. We know the aspect of the daisy plant, of the dandelion, and of the London pride as seen from above, but are not familiar with the side view of such structures, and all plants with the top view of which we are familiar, and which are consequently to us horizontal ornaments, are of a radiate character. On the contrary, all plants with the side view of which we are familiar have but their halves corresponding, and in many cases the halves are not

exactly alike, setting aside the influences of disturbing causes.

50. When the leaves are opposite (see section 10, Chap. VI.) the halves of the structure exactly correspond, when seen laterally, if undisturbed; and the same is the case if the leaves are disposed in rings (Chap. VI., s. 10). But when the parts have a spiral exodus, the halves cannot be precisely alike; yet even here nature seems to make an effort to secure this vertical symmetry, which is pleasing in objects having an upward tendency.

51. In the case of the flower it is similar. If the halves only are alike, the blossom is so situated that it becomes a vertical ornament, as in the violet (fig. 122),

Fig. 122.

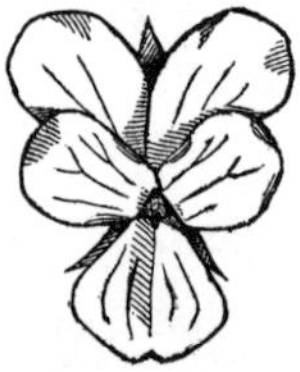

where the head nods, the mimulus, sage, dead-nettles, eye-bright, lavender, and mother-of-thousands (*Saxifraga sarmentosa*), Plate XIII., fig. 1, where the flowers proceed laterally from the stem; but if the flower consists of similar radiating parts (Plate XIII., fig. 2), it occupies a horizontal position, as the rose, sweet-william, pink, and buttercup. Yet we occasionally find flowers formed of similar radiating parts placed vertically; but to this there is less objection than to a bi-symmetrical flower being placed horizontally. It will be said that flowers of a bi-symmetrical character do occur horizontally, and the candy-tuft (Fig. 123), hemlock, and many umbel-

FIG. 123.

liferous plants will be adduced as examples; for in these cases the halves of the flower only correspond, and these do enjoy a horizontal position. Circumstances, it is said, alter cases, and this is true in this instance at least; for here the flower is not solitary, but the blossoms are so arranged that the symmetrical character is restored by their disposition, as the smaller halves of the flower are, in all cases, directed towards the centre of the circle (fig. 124). We learn, then,

FIG. 124.

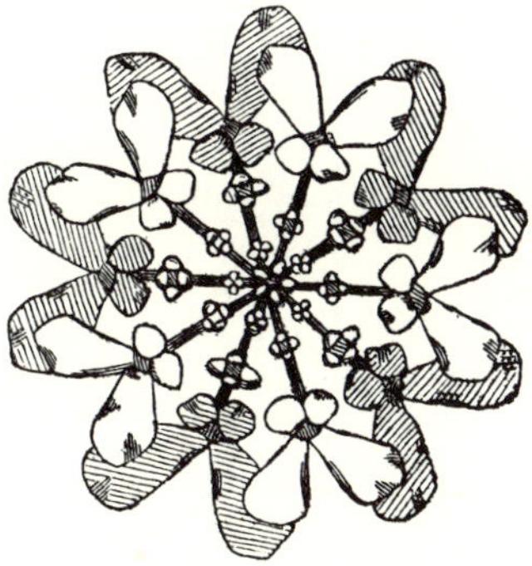

that in the case of those plants which appear as horizontal ornaments we have a radiate structure, and in the case of those plants which appear as vertical objects we have a bi-symmetrical structure. Applying these principles to the case of conventional ornaments, we see that while a pattern constructed on such a form as is represented in fig. 57 is adapted, in its present arrangement, for a wall-decoration, it would be most objectionable as a floor-pattern; for in this position it would of necessity be wrong way upwards to a large portion of the occupants of the room. Yet such patterns are fabricated and used, bad as they are; and when they are purchased and taken home a long discussion ensues as to how the pattern is to run—whether to the fire or from the fire; and all this arises out of the pattern being unsuited to its position. We have seen that, in certain cases, flowers of a bi-symmetrical character have a horizontal position, so it is not necessary that all floor-patterns be made up of squares and circles. Even the form represented in fig. 57, which we have just been considering, is admirably fitted for a horizontal position; but then the arrangement must be that represented in fig. 58, which is pronounced or emphasized equally in different directions.*

52. The arrangement of the diamond or lozenge form, given in fig. 53, is perfectly suited for a wall or vertical surface, while the dispositions of this form, given in figs. 50, 51, 52, and 54, are better suited for floors or horizontal surfaces; and the same thing occurs with the form set forth by fig. 125, which is perfectly suited to a vertical surface when erect, as here drawn, or when by inversion it is changed into

* A carpet pattern of a very simple character founded upon this basis was designed by the author and manufactured by Messrs. Jackson and Graham some time since, and a wall paper to match, on the basis of fig. 64, has recently been cut by Messrs. Scott, Cuthbertson, and Co.

a scale pattern (fig. 126); while, if it is required as a floor-decoration, it must be arranged in the manner represented in fig. 127.

Fig. 125.

53. With stair-carpeting the case is somewhat different; for here the form and arrangement given in fig. 57 would not be so offensive and inappropriate as it

Fig. 126.

would upon an ordinary floor; but even here, immediately the top stair is ascended, such a pattern would be very objectionable. Hence the treatment should be similar to that of the common floor, or the pattern should be accented in two directions only (figs. 128, 129).

54. We have seen that in the case of plants those which are seen as horizontal objects have a radiating

Fig. 127.

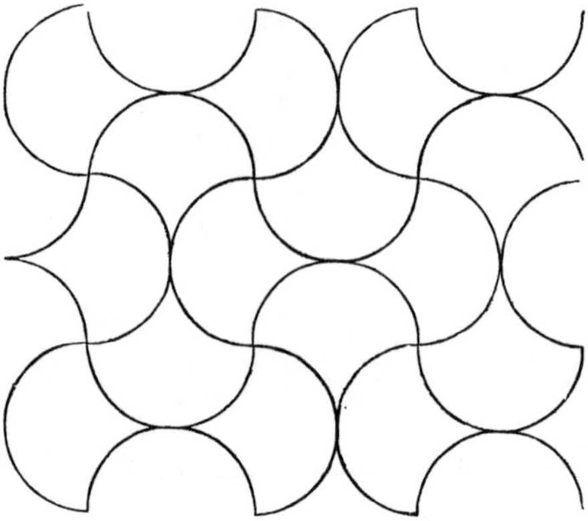

character, while those which appear as vertical objects

Fig. 128.

Fig. 129.

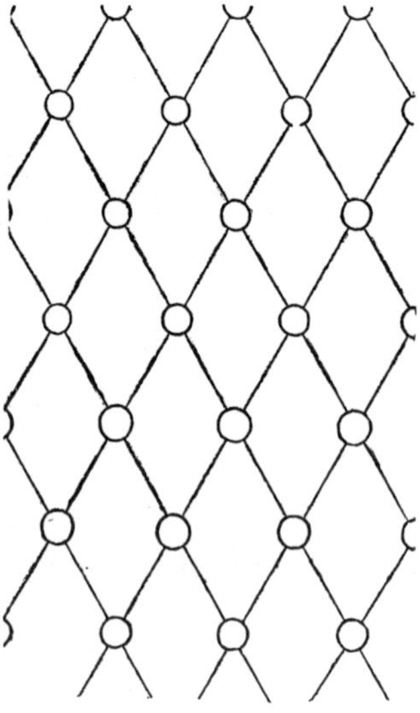

are of a bi-symmetrical character; in conjunction with this one circumstance should be considered, which is, as vegetable forms are of a flexible or elastic nature, it is impossible for a radiating ornament to retain a circular form if placed vertically, owing to the influence of attraction upon plastic matter.

55. For the purpose of illustration, we shall compare a plant, with its circular system of leaves, to whalebone radii terminated in knobs proceeding from a fixed centre, fig. 130. The centre in our figure corresponds

Fig. 130.

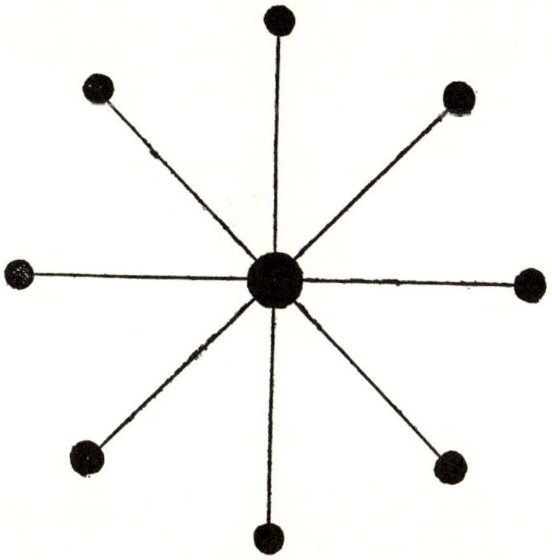

with the short stem of the plant from which the leaves proceed, the whalebone rays, with the leaf-stalks and the knobs on the extremities of these rays, with the leaves; when this is sustained horizontally, and viewed from above, it appears as a regular star; for the depression of the rays by the attractive influence of the earth, to whatever extent it may take place, is unobservable in the position from which we are viewing it; or, to be more accurate in expression, the character of the ornament is not altered, for all the rays are of the same length, and each appears as a right line; but let this arrangement be placed vertically, and viewed as a wall-decoration, and its character is seen to be altered, for it no longer consists of members having the same direction, but of parts so arranged that the whole is now of a bi-symmetrical character (fig. 131), and this one fact seems to give

FIG. 131.

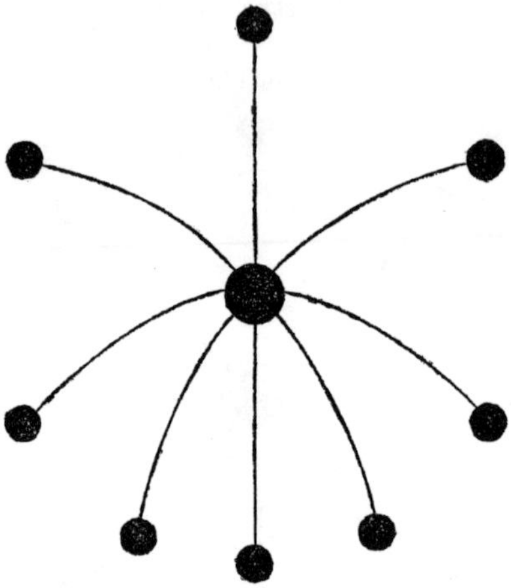

much consistency to many of those forms which we have used as illustrations of repetition with variation (see Plate IX.).

56. In relation to adaptation one other hint is given us by nature through the medium of plants, which is, that the portions of a composition may be removed from their normal stations if the conditions are so altered as to render this desirable. This can be illustrated by the ivy, where the leaves are generated upon the spiral principle, and are so protruded that the sixth leaf falls over the first, and two revolutions of the spiral thread round the stem are made in encountering the fine leaves which form the unit of

the spiral series (Chap. VI., s. 13); the arrangement, indeed, is precisely like that of the oak represented in figs. 68, 69. This is the normal arrangement of the leafage of the ivy, and the disposition which occurs in all cases where an equal supply of air and light is received on all sides, but no sooner is the plant brought in contact with a wall than it completely alters the arrangement of its leafage to the alternate disposition (Chap. VI., s. 12); and even this is so modified that the leaves are all turned towards the spectator, and thus the most beautiful arrangement of the parts possible under the circumstances is brought about.

57. A similar departure from the normal law frequently occurs in the jasmine, which has its leaves arranged in crossing pairs (Chap. VI., s. 10). We

FIG. 132.

here find that when the plant is brought in close contact with a wall the crossing arrangement is departed from, and the leaves pass only to the right and left (fig. 132). In the case of the ivy, one leaf out of five, at least, would come in contact with the wall and be cramped and injured, and consequently of an unsightly character, were it not for this departure from rule; and in the case of the jasmine, one leaf out of four would be destroyed; but by this departure from the normal habit of growth the leaves are preserved from injury, and a new disposition of the parts is brought about which secures beauty.

58. An instance of a somewhat similar character, having a more special reference to the relation of wall and floor-patterns, or the manner in which a pattern suited to the former position should be modified in order to be appropriate to the latter position, we have noticed occurring especially in the goosegrass, where the leaves are arranged in rings in the manner represented in fig. 61. This plant has feeble stems, yet it is destined to an erect position; for the angles of the stem and the midribs of the leaves are each furnished with hooks which by their graspings sustain the plant in an erect position. When thus vertical, the leaves are given off from the stem in all directions, at equal angles, a condition shown in fig. 61, to which we have just made reference, but if by any accident the seed springs up, and the plant develops itself where there is nothing on which it may lean, the stem reclines upon the earth, and develops in an onward instead of an upward direction. Here, however, is the circumstance to be noted: the direction of the leaves is now changed, for instead of their proceeding from the stem in such

a manner that they all form with it equal angles, some leaves are so raised, and others so depressed (fig. 133), that we have an arrangement which appears

Fig. 133.

in top view as a series of rings of leaves connected by a stem (fig. 134.) What do these instances teach

Fig. 134.

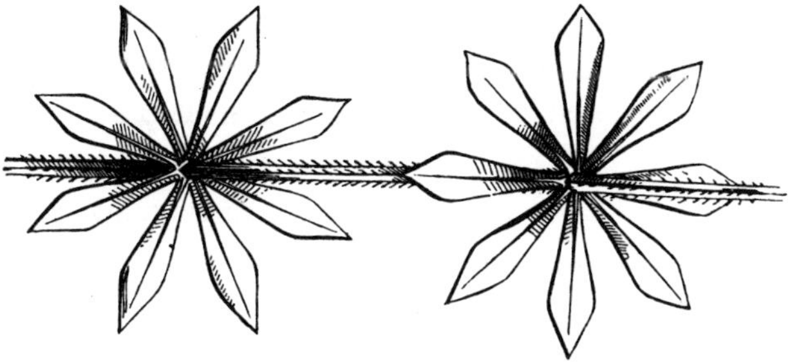

us? Is it not that fitness must precede all other considerations, and that we must apply ourselves to the task of discovering the most befitting form, or appropriate condition of an object or surface, and after this seek to apply such enrichments to it as seem desirable and in no way violate the principle of consistency? We now see the teachings of nature upon the subject of adaptation, and the application of these teachings to the case of decoration; yet even with this knowledge, and an acquaintance with the practice of those ages in which art has flourished, which has in all cases been in unison with the principles we are advocating, we cannot hope that any great advances will be made, so long as articles of furniture are chosen in the manner in which they now are. If a house is to be furnished, or even a room, the purchaser selects first a pretty paper for the wall, then a pretty carpet for the floor, then pretty curtains for the windows, a pretty cover for the table, pretty chairs and tables, and as many pretty things as are deemed necessary, without ever thinking whether all these pretty things will make a pretty whole. The carpet may be good as a carpet, but may combine with the wall anything but harmoniously; the table-cover may be well chosen as a solitary object, but may prove to be the opposite of what it ought when brought into society; and both the curtains for the windows and the covers of the chairs may be offensive when in juxtaposition with other objects: so long, we say, as we proceed in this manner, little advance can be looked for; we must select those objects which are in all respects best adapted to the circumstances in which they are to be placed, and this can only be done where knowledge is possessed.

CHAPTER XII.

THE POWER OF ORNAMENT TO EXPRESS FEELINGS AND IDEAS.

1. A DRINKING vessel preserved in the South Kensington Museum is beautified by an ornament which sets forth the Chinese notion of an eclipse; a golden ball represents the sphere about to be lost in darkness, a number of clouds float in mid air, and a monster dragon advances with the view of swallowing the golden orb; here we have a subject—a thought—an idea, shrouded in ornament or set forth by decorative forms,* and that in a manner alike beautiful and impressive, for the dragon is all that could be desired, being vigorous, energetic, grotesque, and ornamental; but the success of the composition, great as it is, does not concern us at present, as our attention must fix itself upon the fact that a *subject* is portrayed by ornamental characters, and upon the agencies employed to bring about this end. (Plate XIV.)

2. In paragraph 14 of Chapter II. we have shown that an image possessing the salient characters of an object will call the object to mind, while the image is

* We could not regard the dragon as a piece of ornament were it a representation of any existing creature, but as it is a mere ornamental animal, we are justified in so doing.

in all minor particulars unlike that which is brought before the mental eye, which truth being recognised by the Chinese artist has enabled him to call forth within us the idea of a dragon by an ornamental creation which, while it is possessed of eyes, legs, body, tongue, and teeth, is yet unlike any creature that either does or can live, yet unmistakeably a dragon, for the salient features in the mental conception of this monster are present, which of themselves suffice to enable the mind to recognise the intended thought; but in this instance the inducing the mind to create to itself a particular creature is not difficult, as the term dragon is applied generically to all animals with a long body continued as a tail, short and powerful limbs, and a head furnished with a large mouth and powerful jaws, as well as by the dragon being a mere fabulous creature. The clouds are depicted in a manner so conventional that unless the significance of the forms used were known it might be difficult to determine what was portrayed; yet in a symbolic system the use of forms of a wholly conventional character is sanctioned by the practice of ages as the counterparts in ornament of natural objects.

3. This illustration makes manifest the possibility of setting forth an idea by decorative forms, and we might bring forward others which would show that a plurality of thoughts can be illustrated by ornament, and that knowledge may be shrined in beautiful forms. We frequently find upon Chinese vessels the earth represented as a plain, the plain in some instances resting on the back of a tortoise, and the conventional clouds floating above; thus this ingenious

1

2

people set forth that which they believe. We know the earth to be spherical, and to be surrounded by clouds and starry worlds, hence it is but right that we set forth our knowledge in ornament as do the Chinese. In order to this, we draw the earth as a globe, or, to be precise, as an oblate spheroid; we mark upon it the great continents and a few longitudinal lines; we now surround it with clouds, using the form furnished by the middle-ages, and place externally the stars. (Plate XV.)

4. Although we see the possibility of expressing definite ideas by the agency of ornamental shapes, yet this has been accomplished by the aid of forms having an acknowledged significance; and pleasing as this mode of expression by the agency of symbols is, yet the power of utterance thus afforded is limited. Orange-blossoms may set forth innocence, and call up reminiscences of Hymen's altar; holly and mistletoe will express Christmas, and evoke thoughts of merriment and joy; the violet will typify humility, leaf-buds spring, a profusion of flowers summer, ripe fruits autumn, leafless branches winter, the trefoil the Trinity, and the quatrefoil the four Evangelists; yet the number of such symbols is few, consequently they afford no great means of expressing ideas.

5. In order that a symbolic art be powerful in utterance it is necessary that the people have knowledge of the purport or significance of the forms used and of the circumstances to which they have reference, for in the absence of this knowledge the decoration, through not being understood, ceases to be impressive; and this is the great drawback to a symbolic system, and

is that which will ever limit its development and retard its adoption, even if nourished by favouring circumstances; but as the most propitious conditions for the healthy growth of such a system have passed away, there is no reason in hoping that symbolism will again prevail. Still, all are agreed as to the desirability of embodying our knowledge in ornamental forms; indeed, this must be done in order that decoration command our sympathies and be satisfying, but it must be accomplished without the aid of symbols. The appeal must be made to that knowledge which is invariably gained by the common-life experience of those addressed, or to the education which such are known to possess. This is the only way in which we can hope to exalt ornament, and to endue it with the power of eliciting the sympathies of the people, and of drawing to itself that patronage and interest necessary to its advancement. We condemn the conduct of the people in not nourishing the growth of ornament, and upbraid them on account of their ignorance of art, saying, that this alone is the cause of the poverty of the harvest of decorative forms.

6. But is it so? Is there no fault on our side? Have we furnished ornaments which, while they are comely in form, also manifest the knowledge of our age? And have we placed our fellow-countrymen under an obligation by furnishing their houses with decorations which will proclaim to generations yet unborn the nature and extent of our discoveries, and which, at the same time, instruct those who now dwell upon them? If we have not done so, can we wonder that art is not cherished more? Do we not rather marvel

that in this respect things are not worse than they are? So long as we repeat ancient forms, or continue to produce designs which in no way set forth our knowledge, we must fail to invest ornament with that power which it ought to possess, and that which will bind it to the people. The mechanical arts are justly set forth; our stoves and bronze vases are well cast; our marble mantel-shelves are well cut and finished; the furniture of our rooms is well put together; the glass shades are well blown; the windows are of large

FIG. 135.

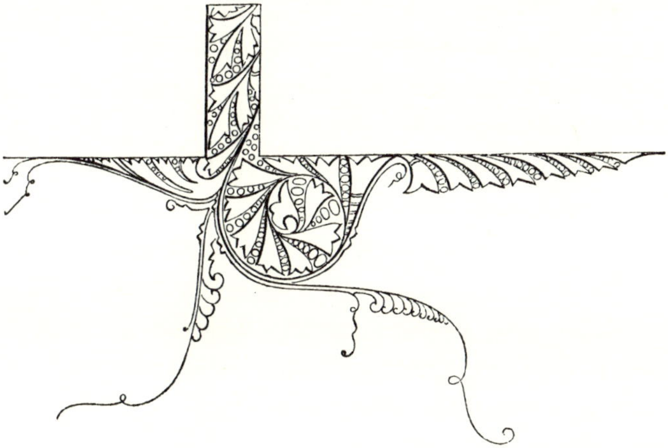

sheets of smooth, bright, transparent glass; the rugs are well made, the carpets well woven, and the wall-papers printed with great exactness,—which all teach us the perfection of our machinery and the greatness of our mechanical skill; and they will also tell, in future ages, of the vastness of our power; but the art will fail to speak of our greatness, powerful in utterance though art be.

7. Fig. 135 will serve to elucidate the manner in

which knowledge can be embodied in ornament and illustrated by it, as it has two members which we may consider as setting forth natural facts. If a ball is thrown against a wall, it will rebound, and that in such a manner that the angle formed by it and the upper portion of the wall as it rebounds, will be coincident with that formed by the ball and the lower portion of the wall, as it first left the hand. If the ball be replaced by a mobile fluid, it will manifest the action of a new influence; for while a portion rebounds, part will run upwards in contact with the vertical surface, being thus drawn from its course and held to the wall by surface attraction. This is well seen at our coasts, where wave after wave advances, and after running up the gravel beach, strikes against the cliff: one portion of the water immediately rebounds, while another portion ascends in contact with the rocky surface. In some places the cliff has been sapped by the ever-rolling waves, where the water may be seen striking against the vertical rock at the back, and then flying off and striking the overhanging surface, which it continues long in contact with while urging its way to the margin of the overhanging rock. Another familiar instance in which surface attraction is made manifest is furnished by an experiment occasionally seen at the tea-table. The spoon, when poised on the edge of the cup in such a manner that the bowl floats on the tea, draws the liquid from its position of rest into an elevated ring in immediate contact with the spoon, which appears to sustain the spoon, and gives a clinging effect, which is applied in Plate XVII., fig. 2, and Plate XXIII., fig. 2.

The circumstances or facts of these examples may be considered as embodied in our figure (fig. 135), where from the spiral a member is given off to the right in a diagonal yet upward direction, which, through coming in contact with a solid surface, rebounds in part and is in part conducted along the horizontal plane. The rebounding of the member, which appears to be urged forward by a vital power, manifests obedience to natural law; and its continuance in contact with the surface yields the same submission to the forces of nature, while the quaint drawing aside of the member given off by a weak force to the left (the weakness of the energy by which it is urged forward being manifested by the fact that it is removed thus easily from its course) shows equal submission to the same influence. This ornament, derived from a high source, (*The Preacher*, illuminated by Mr. Owen Jones) in order to check all doubt concerning its beauty while yielding obedience to decorative laws, may justly be said to embody two natural facts with which we have acquainted ourselves, and these it manifests without the aid of forms having an acknowledged significance.

8. This may be thought to be an exceptional case, but it is not such. Doubtless any thought may be set forth by ornamental forms; but in order to this an inquiry into the facts and circumstances which go to make up the mental conception of that which is to be set forth must be instituted and carefully conducted. What goes to form our conception of evening? Not one fact alone, but a number which we have become familiar with throughout life: it is these which we have to ascertain and discover. Some days

pass away, and are succeeded by night, without our being impressed with the period of twilight, while at other times we *feel* that it is evening. What is it that causes us thus to feel? And what do we now see that we do not meet with at other times? The aërial vault has a hue peculiar to this period of day—a fact which can never fully be appreciated till an attempt is made at mixing a flat tint which exactly represents it. There is the brightening "star of the evening," which shines forth long before its comrades less brilliant than itself. Measure with exactness the means necessary to represent this shining sphere which has a colour that cannot be told, and brilliance which can only be represented by heightening the contrast between the star and the sky. The herbage partakes of the influence, and alters the position of its leaves, which for the most part rise and form a more acute angle with the upper portion of the stem. And the majority of flowers close, while others open, as the evening primrose.* The colours are also altered; blue becomes lighter, till in the later dusk it is impossible to tell a surface washed with cobalt in its intensity from a sheet of white paper. Orange and red become darker, and finally disappear in black. The bright orange flower of the common nasturtium (almost orange-chrome in colour) at the period of the later dusk presents the mahogany or rich chestnut colour of the dark flower of the same

* This they do at various periods of the evening, which usually correspond with the hours of opening in the morning: thus, if a flower opens at six o'clock in the morning, it usually closes at six in the afternoon; if it opens at eight in the morning it closes at eight in the evening. See the "Floral Clock," given in the Appendix.

PL. XVI.

1

2

species,* while yellow is least changed under these circumstances. Greens become darker, objects are blended and confused, and soft indistinctness prevails. Besides these characters, there are owls, the moon, mushrooms, bats, and large-eyed monsters, all of which contribute to the mental conception of this period of day.

9. Having ascertained what goes to make up our mental idea of evening, it is not difficult to set forth the thought in ornament, or to create a decorative scheme which shall so impress the mind as to cause it to originate to itself the sense of evening. But in order to this it is not necessary that we avail ourselves of all the elements which are associated with evening, for a few salient features will usually suffice. All that is necessary is that we originate a series of conventional forms which shall call to mind the salient features in our conception, and the result is gained; but the success of the effect will rest upon the character of the ornament, the arrangement of the composition, and the right perception of the characteristic features. Plate XVI. is a hasty sketch in which we have endeavoured to convey the thought of the evening star.

10. We next offer an illustration of a "Hobgoblin," —a creature with eyes as large as tea-saucers, resident in Pad-foot-lane. (Plate XVII., fig. 1.) It will be found that we associate with our conception of evening

* The flowers of the common varieties of the nasturtium, such as we find in cottagers' gardens, are in colour either lemon-yellow, orange, or of a dark and rich chestnut-maroon; the two latter are alluded to above.

the existence of creatures with luminous eyes; for insects generally impress by their eyes and wings, and although perfectly familiar with the organs of vision of these creatures, we are frequently altogether ignorant of the character of the mouth. Hobgoblins are associated with evening, hence it is necessary to link them with our conception of this portion of the day and to make them terrific through the magnitude of the eyes. In Plate XVIII., fig. 1, another form of monster is given.

11. It should be remembered that a play upon other forms is suggested by plants: for various resemblances to animal forms appear in the Orchid flowers; flies are seen in the fly-ophrys, bees in the bee-ophrys, musquitoes are borne by the musquito-flower of the Peruvians; the butterfly is found in the butterfly-plant of Trinidad; pelicans in the pelican-flower, which is a kind of lady's-slipper; the skin of the tiger in the tigerlike stanhopea; that of the leopard in the leopardlike bolbophyllum. The lynx-like flower of Hernandez is so called from its lynx-like eyes and teeth. Then we have the military-orchis and the monkey-orchis, the lizard-orchis, the lady's-slipper, dragon-flies, moths, swans, doves, bulls, serpents'-heads, and frogs in various other flowers of this extensive order; hence, in our practice of playing upon other forms we are but following a hint given by nature in some of her most beautiful manifestations.

12. Our sketch of the Hobgoblin may be appropriately followed by one of evening in which this creature is present. (Plate XIX.)

PL. XIX.

1

2

13. The most simple agency by which the thought of ascension can be called to mind is by the arrow with its apex directed upwards, and of descension by an arrow inverted; but the ideas evoked by such a use of this simple form are necessarily indefinite. In order that a sense of ascension and descension be more perfectly awakened a more complicated mode of expression must be adopted, which shall be an embodiment of occurrences or effects associated by the mind with these acts. An umbrella, moved through the air with considerable velocity, in such a manner that the handle follows, has its web pressed closely against the stick. A bird in rapid flight has its feathers pressed against the body. A parachute during ascent is closed. Should the umbrella be pulled downwards through the air it will be forced open; a bird when descending has its wings uplifted; and the parachute is spread—hence these bodies, when in these particular conditions, tend to induce the mind to originate to itself the thought of ascension or descension, or progression and retrogression, as they are directed perpendicularly or horizontally. As illustrations of what is necessary in order to the awaking the thought of ascension and descension, we give the following illustrations. (Plates XX., XXI., XXII.)

14. Our ideas of ascension and descension are derived almost exclusively from the observation of bodies rising and falling through the atmosphere; we may have observed objects rise and sink in water or other liquids, but from bodies moving through the air our chief sense of these acts is gained. The air is fluid

matter, having density and weight, and offering resistance to all bodies moving through it; hence if an object moves through it, it is pressed upon, as the air has not the motion of the moving body; and if the object be flexible or elastic its form is altered, or the direction of its parts changed by the resistance offered. The nature of the resistance must be carefully considered, and the manner in which it is offered. If the medium, through which the body moves were more dense, the resistance offered would be greater; thus motion through water might be set forth as well as motion through air, but the want of familiarity with the aspect of bodies when moving through this medium might debar it from being appreciated. The effect of pressure on bodies of varied degrees of elasticity and extensibility, both free and fixed, should be most carefully and constantly studied. Some little time back we were much instructed on this subject by a series of experiments performed by Professor Tyndall, at the Royal Institution, illustrative of his views of glaciers; he caused semi-plastic matter (pipe-clay reduced to the consistency of thick batter by water), to flow down planes of varying declivity, having coloured lines and circles impressed upon it, which underwent remarkable changes of form as they descended and were acted upon by varying pressure. The rings resulting from bubbles of phosphuretted hydrogen gas being introduced into the atmosphere, when acted upon by gentle currents of air, manifest the action of force very readily. The influence of matter in motion upon matter, whether moving or fixed, is at all times worthy of study, for by observing the aspects of matter when acted upon by various influences, and diligently inquiring into the nature of the mental conception of facts and occurrences, it will be found possible to express feelings and ideas by ornaments without the aid of symbolic forms.

CHAPTER XIII.

PRINCIPLES COMMON TO ORNAMENT.

1. THERE are a number of principles which, while very important in relation to ornament, are yet of minor interest when brought in comparison with those which we have already discussed; yet while of small moment, when thus compared with the most important laws of decoration, they are not the less worthy of careful consideration. The key-stone is but small in relation to the arch, yet very important; the eye is but small in relation to the body, yet of incalculable worth; and just as the arch could not stand without the key-stone, and the body suffers a great want if the eye be absent, so these principles, while yet relatively small, are so momentous that if they be neglected the whole composition is marred.

2. In setting about the production of a design, the first thing to be considered is the fitly dividing the space or form to be ornamented into primary divisions or shapes. This may be accomplished by bands, spaces, interlacing-lines, or in ways which can only be fully comprehended by a careful inspection of the works of great masters; and in such investigations the student may profitably spend much time.

3. In the primary divisions, the utmost attention

must be paid to proportion (Chap. IX.) and consistency. If the shape to be divided is of a geometrical character, the mode of division set forth in Chap. V., s. 9, may be advantageously employed; but should it be a vase, then the primary lines of the ornament must tend to emphasize the beautiful quality of the object, and not to destroy the form, for no decoration is tolerable which militates against the general beauty of the shape.

4. We have seen that in the selection of a form the utmost regard must be paid to utility, and that the usefulness of an object or vessel must never be impaired by its decoration (Chap. XI., s. 3); so in relation to the general division of the form, it must be such as will tend to impress the use as well as the beauty of the object or surface decorated.

5. After the division into primary forms, the secondary shapes and lines may be cared for; but to the excellence of the primary forms too much attention cannot be given. In conjunction with the secondary forms, the masses or spots in the ornamental scheme, which may be rendered more or less prominent by magnitude, or strength of colour, or the relative lightness or depth of the tint, may be regarded; and these secondary masses are also worthy of very special consideration.

6. After this the detail may be thought out; and while an excess of merit cannot in any place occur, yet the primary setting out is of the greatest importance, then the secondary forms, then the detail; for at a distance the primary forms can alone be seen, while in order that the secondary forms be perceived a

nearer view must be had, and only upon closer inspection is the detail visible, as with refinement there is always softness and delicacy: those decorative schemes which have all their parts in such harsh contrast that the minutiæ can be seen while the spectator is far from the object, are cut off, by their vulgarity, from commendation.

7. When we see a tree at a distance its general form strikes us, but nothing more. It may have the form of an egg, as a well-grown horse-chestnut, or of a spindle, or lance, as the tall poplar, or of an umbrella, as some cedars; but whatever its shape be, nothing strikes us but the general form when it is thus viewed.

8. Upon approaching nearer to the tree we observe a general division into large masses—masses which, in tall trees, are elongated, and in broad trees are depressed; but it is only by our being in close proximity to the tree that we discover the form of the leaf and the general detail. This is the teaching of the plant upon the subject, and the student cannot do better than follow the dictates of nature.

9. Effects of varied characters can be produced by the adoption of distinctive treatments and materials. The lavish use of elongated pointed forms and of sharp lines of white or gold will give a crisp and thorny aspect, whereas larger forms of a less pointed character will give breadth, which is frequently desirable; this difference occurs between the foliage of the holly bush and fig-tree. These things are mentioned in order that the attention of the student be awakened to the means afforded of varying the effect by the alteration in character of the material used,

for all power is of value; but with the increase of power there is also a proportionate increase of responsibility—hence care must be used in fitly appropriating the ornament to the surrounding circumstances.

10. Mr. Owen Jones justly says, that as a composition to be satisfactory in its colouring must have the presence of the primary colours, red, blue, and yellow, in their intensity or in combination, so must a composition, to be perfect in form, have the presence of the three lines—the straight line, the diagonal, and the curve. Thus fig. 136 is not satisfactory, for it carries the mind up, down, and across. Fig. 137 is not satisfactory, for it still leads the mind outwards; but fig. 138 gives rest, as the circle fixes the attention and binds it to a focus.

FIG. 136.

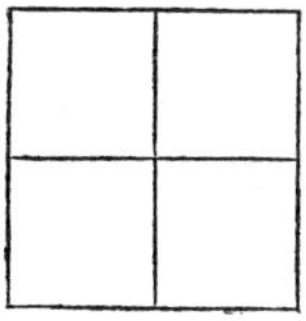

FIG. 137.

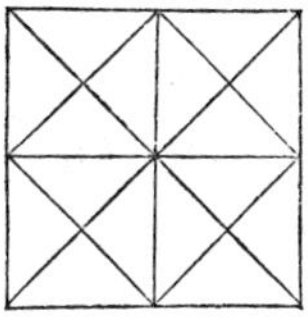

FIG. 138.

FIG. 139.

FIG. 140.

FIG. 141.

11. It must not be understood that all ornamental compositions are to have the basis given in fig. 138; for this is by no means desirable; yet this principle can be present while much altered in aspect. Figs.

139, 140, 141, are so many embodiments of the spirit of the statement, while in fig. 140 the most prominent set of lines is represented by spaces; thus we see that the principle can be embodied while the form is yet modified; and this alteration of character admits of endless variation, while the spirit of the proposition is yet felt and preserved. See Plate XXIII., fig. 1.

12. In a composition an amount of attention should be given to space as well as form; indeed in some cases it is as agreeable to see large and small spaces as it is to observe large and small forms, which are very necessary, as we gather from section 6 of this Chapter. But while it is necessary that some attention be paid to a judicious alternation of space

FIG. 142.

and form, too much attention must not be given to the ground, as the beauty of the ornament must never be sacrificed to the aspect of the space. In fig. 142

FIG. 143.

FIG. 144.

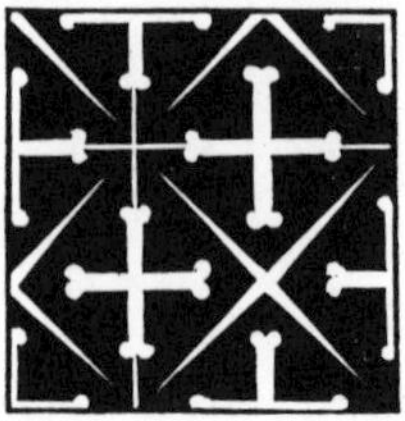

we have the most simple illustration of that class of pattern in which the ground and figure are both

1

2

Day & Son, Lith^rs to the Queen

PL. XXIV.

ornamental and are yet dissimilar; and in figs. 143, 144, 145, and 146, the same principle is applied in ornament, while fig. 147 illustrates the principle of having

FIG. 147.

FIG. 145.

FIG. 146.

form and ground similar, which we have carried into ornament in figs. 148, 149, 150, and 151, but this can be rarely done without the ornament suffering.

FIG. 148.

FIG. 149.

FIG. 150.

FIG. 151.

13. A class of decorations of a distinctive character arise from the ground being wholly covered with ornament; but as the construction of this form of pattern (Plate XXIV.) is attended with difficulty, the shapes frequently becoming cramped and imperfect unless

great care is exercised, it should be the last class of pattern attempted by the student. This form of composition has its analogue in water crystallized on glass, as the frost on the window, rather than in any portion of the plant.

14. In ornament, as in plants, foliage should never spring from branches growing in opposite or antagonistic directions, while yet united, as in fig. 152, for

FIG. 152.

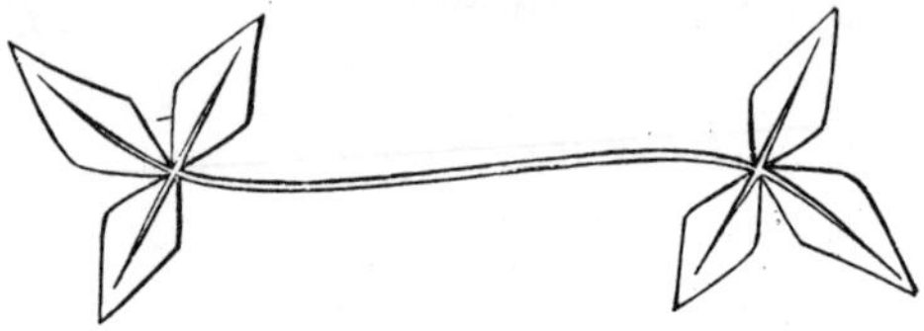

this is in all cases objectionable. The root, in the case of the plant, grows in a manner contrary to the stem, for the one penetrates the earth or strikes downwards, while the other ascends into light and air. But here the leafage only occurs on the ascending members and not on the roots; besides which, the roots are hidden in the earth in almost all cases; hence the teachings of plants seem to be in antagonism to the foliaceous growth in opposite directions as well as the invariable practice of all the great masters; hence it must ever be avoided. If an arrangement similar to fig. 152, be desirable, the line

FIG. 153.

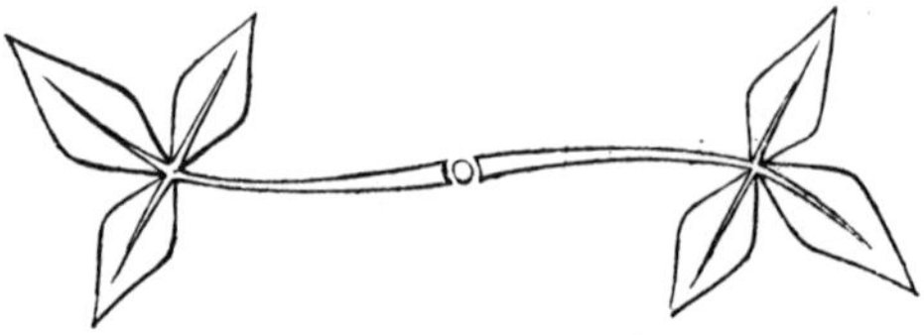

should be broken in the centre, as in fig. 153, for this breach of continuity alters the character of the composition, it being now resolved into a mere grouping of sprays.

15. Besides the principles which we have already laid down, there are many other little things which demand attention. An arrangement of parts, for instance, in an ornamental composition such as tends to suggest the human countenance, or any monster, should be scrupulously avoided. Fig. 154 represents

Fig. 154.

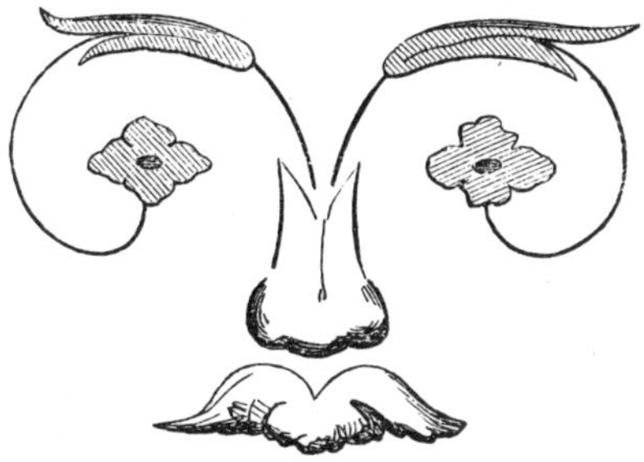

one of four similar appearances in the case of one wall-paper pattern; and in Plate XXV. a lady of the character represented in the lower figure, with a handkerchief over her shoulders, will be distinctly seen if the ornament be placed at a distance from the spectator. Little things of this character are so numerous that the student must trace them out for himself; but before taking a farewell of our reader, we should say a word upon the starting-point in a design and the turning the corner, which to the young student are both difficulties.

16. That starting is the best which is most simple and easy. As in literature, a forced commencement is

always objectionable, and a natural and easy beginning is that which we like, so in ornament that which is easy is agreeable.

17. A simple braiding of the stalks is in some cases very acceptable, as in Plate XXVI., figs. 1, 2, 3; or a mere ornamental treatment of the end of the branch (the heel), Plate XXVI., fig. 4, such as it seems to suggest, Plate XXVI., figs. 5 to 17; or the growing out of the ornament from a line which forms part of the basework of the composition, as in Plate XXIII., fig. 1. The starting with the abrupt end of the stalk is in some cases not offensive, as in Plate II., fig. 7, and the springing the chief lines

Fig. 155.

Fig. 156.

Fig. 157.

Fig. 158.

from little simple forms, such as figs. 155, 156, 157, and 158, is in some cases desirable; but if any principle

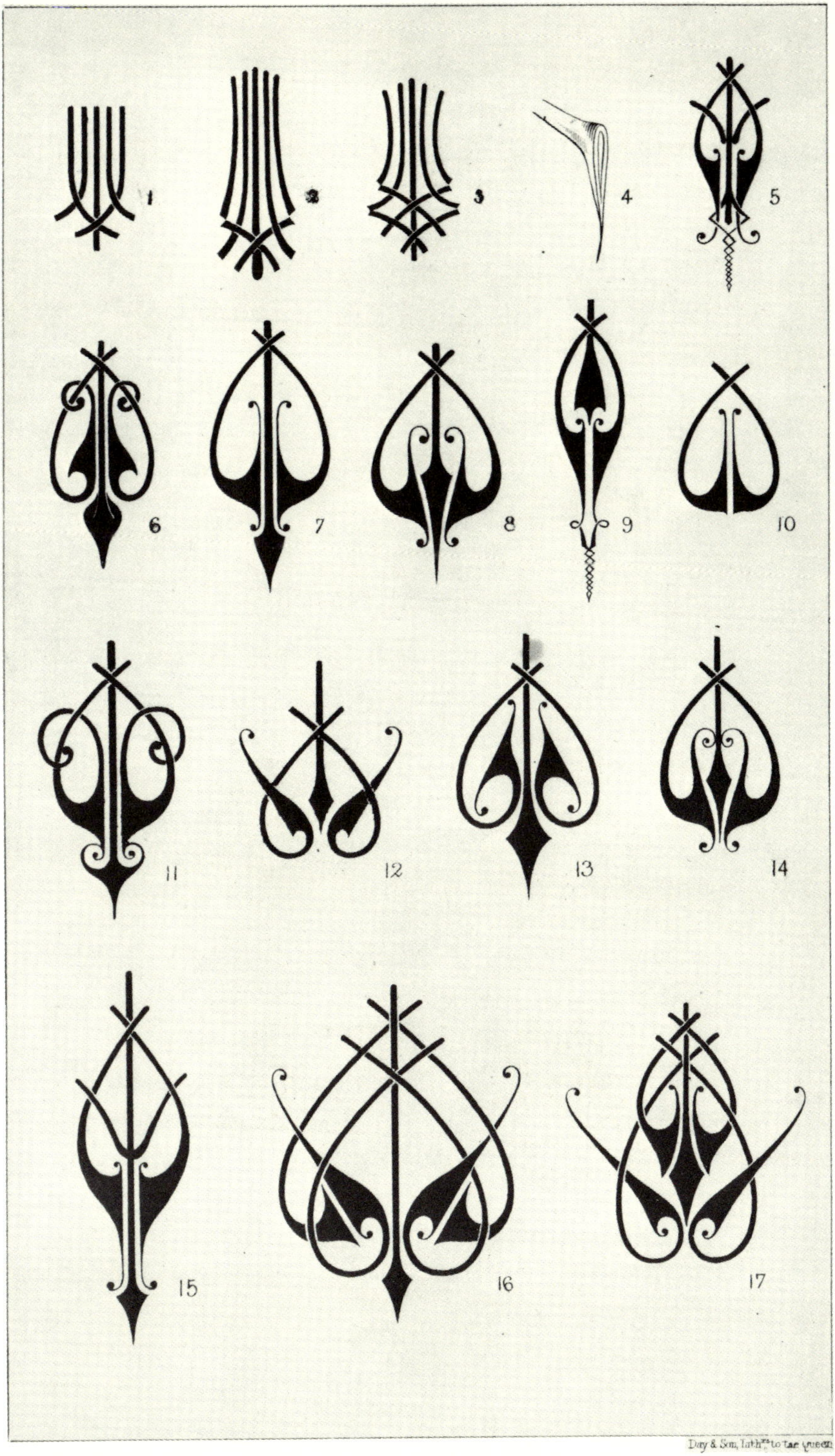

Day & Son, Lithrs. to the Queen

of starting is to be especially commended, it is that which is least obtrusive, and indeed, is least like a formal commencement.

18. In relation to the parts of the design it should be noticed that the members may in some cases be to an extent dissevered from the parent stem with advantage, as in figs. 159 and 160; but in this case great care

FIG. 159.

FIG. 160.

must be used in order so to arrange the parts that their origin can be traced, and that they do not appear as scattered fragments, which are very objectionable; but with care this mode of treatment is in some cases very valuable. (Fig. 78 and Plate XVIII., fig. 2.)

19. What we have said of the starting-point will in part apply to corners, for an easy and simple mode of treatment is here also necessary. Plate XXVII. is introduced with the view of setting forth two modes of treating the corner; in the one case the angle is treated as such, and in the other the ornament merely flows round; but in this, as in other difficulties, reference must be had to great works, by which a true knowledge of the best modes of overcoming the difficulty will be gained.

20. Another point which demands notice in this

summary of minor principles is that two centres may in some cases be employed in a composition, and that in two distinct ways. Thus in the one case the parts may spring from one centre and meet in another (fig. 162), a condition of things found in fig. 1, Plate XXIII. In the other case, while the parts bear a relation to one line as a centre (the line *a b* in the figure, fig. 162), they also bear a special relation to another line as a centre (the line *c d*; see also Plate XXVIII. and Plate XV., fig. 2).

FIG. 161.

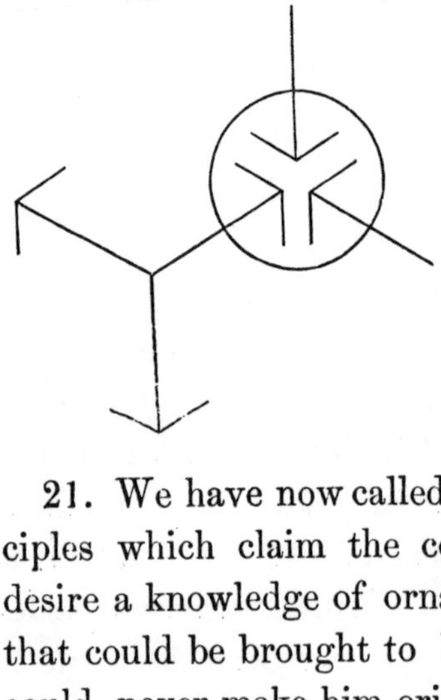

FIG. 162.

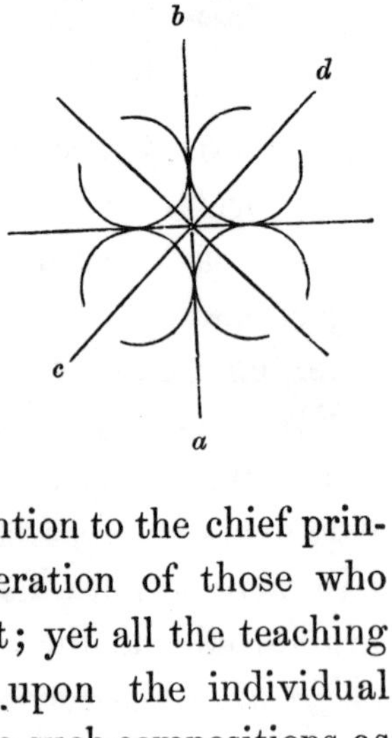

21. We have now called attention to the chief principles which claim the consideration of those who desire a knowledge of ornament; yet all the teaching that could be brought to bear upon the individual could never make him originate such compositions as by their beauty delight, if separated from the most persevering efforts and constant feeling after beauty. But with diligent inquiry, persevering application, and long-continued thought, the greatest difficulties may be surmounted. The designer's mind must be like the

vital force of the plant, ever developing itself into forms of beauty, yet while thus free to produce, still in all cases governed by unalterable laws; and in the action of the mind being controlled by rules we rejoice, and not mourn—"for he who is obedient to the great parent laws of nature is the freest man."

> "He is a freeman whom the truth makes free,
> And all are slaves beside."

APPENDIX,

CONSISTING OF

The "Floral Clock," or the Hours of the Day at which Flowers open.

The Ornithological Clock, or the Hours at which Birds rise.

The Characteristic Flowers of the Months.

Other Characters of the Months.

The Characteristic Plants of various Soils and Situations.

The Characteristic Plants of various Countries,

THE FLORAL CLOCK.

The Floral Clock, *Horologium Floræ,* resulted from the observation by Linnæus of the hours at which plants expand and close their blossoms. Linnæus found that the flowers open with such regularity that the hour of the day is marked by their unfolding; and he likewise observed that the closing of the blossom indicated the corresponding hour in the afternoon. Thus, if the flower opens at six o'clock, A.M., it closes at six o'clock, P.M. : this, however, is not invariably the case, and must not be relied on, while the period of opening is much more certain. Indeed, the hour of closing is tolerably certain also; but it is not regularly twelve hours from the time of opening. The Floral Clock suggested by Linnæus, yet in a much more extended form than that given by him, we subjoin, as it may be of value to the ornamentist and artist in affording a means of marking the periods of day.

	Hours at which the flowers open.	
Major convolvulus. (*Ipomœa purpurea*)	2	a.m.
Ipomœa nil	3 to 4	„
Bindweed. (*Calystegia Sepium*)	3 to 4	„
Yellow hawkweed. (*Tolpis barbata*)	4	„

	Hours at which the flowers open.
Yellow goat's-beard, or meadow salsify.	
(*Tragopogon pratense*)	4 to 5 a.m.
Matricaria suaveolens	4 to 5 ,,
Buttercups.	
(*Ranunculus*)	5 ,,
Wild succory.	
(*Cichorium Intybus*)	5 ,,
Papaver nudicaule	5 ,,
Dandelion.	
(*Taraxacum Dens-leonis*)	5 to 6 ,,
Minor convolvulus.	
(*Convolvulus tricolor*)	5 to 6 ,,
Common lapsane, or nipple-wort.	
(*Lapsana communis*)	5 to 6 ,,
Sow-thistle, or Milk-thistle.	
(*Sonchus oleraceus*)	6 ,,
Spotted cat's-ear.	
(*Hypochœris maculata*)	6 ,,
Convolvulus siculus	6 ,,
Water lilies.	
(*Nymphæa*)	7 ,,
Wall or ivy-leaved lettuce.	
(*Lactuca muralis*)	7 ,,
Small Cape marigold	7 ,,
Bearded fig-marigold.	
(*Mesembryanthemum barbatum*) . .	7 to 8 ,,
Venus's looking-glass.	
(*Campanula* or *Specularia Speculum*)	7 to 8 ,,
Cucumis anguri	7 to 8 ,,
Yellow mouse-ear	8 ,,
Scarlet pimpernel.	
(*Anagallis arvensis*)	8 ,,
Nolana prostrata	8 to 9 ,,
Field marigold.	
(*Calendula arvensis*)	9 ,,
Sandworts.	
(*Arenaria*)	9 to 10 ,,
Knotted fig-marigold	10 to 11 ,,
Green colewort	10 to 11 ,,
Alpine dandelion	10 to 11 ,,
Common star of Bethlehem, or Lady Eleven-o'clock.	
(*Ornithogalum umbellatum*) . . .	11 ,,

	Hours at which the flowers open.	
Many fig-marigolds. (*Mesembryanthemum*)	12	noon.
Common passion-flower. (*Passiflora cœrulea*)	12	„
Carnation. (*Dianthus*)	1	p.m.
Afternoon squill.	2	„
Pyrethrum corymbosum	2	„
Purple hawkweed. (*Barkhausia rubra*)	4	„
Night-flowering catchfly. (*Silene noctiflora*)	5 to 6	„
Evening primroses. (*Œnothera*)	6	„
Marvel of Peru. (*Mirabilis*)	6 to 7	„
White Lychnis. (*Lychnis vespertina*)	7	„
Cereus grandiflorus	7 to 8	„
Mesembryanthemum noctiflorum	7 to 8	„
Œnothera tetraptera	7 to 8	„
Œnothera suaveolens	7 to 8	„
Night-flowering cereus	8 to 9	„

ORNITHOLOGICAL CLOCK.

The Floral Clock may be appropriately followed by the Ornithological Clock, constructed by a German woodman. This clock has been formed by noting the hours at which birds wake and sing in the morning. "The signal is given by the chaffinch, the earliest riser among all the feathered tribes. Its song precedes the dawn, and is heard in summer from half-past 1 to 2 o'clock, a.m. Next, from 2 to half-past 3, comes the blackcap (*Sylvia atricapilla*), whose warblings would equal those of the nightingale if they were not so very short. From half-past 2 to 3 o'clock the quail is heard; from 3 to half-past 3 the hedge-sparrow; then from half-past 3 to 4 we have the blackbird, the mocking-bird of our climate, which imitates all tunes so well that all the blackbirds of a French canton were made to sing the Marseillaise Hymn by letting loose a blackbird which had been taught that tune. From 4 to half-past the lark pours forth its melodies; from half-past 4 to 5 the black-headed titmouse is heard; lastly, from 5 to half-past the sparrow awakes and begins to chirp. To throw this into a tabular form, for the convenience of reference, it is as follows:—

Chaffinch	1.30	to	2	a.m.
Blackcap—(*Sylvia atricapilla*)	2	to	3.30	,,
Quail	2.30	to	3	,,
Hedge-sparrow	3	to	3.30	,,
Blackbird (mocking-bird)	3.30	to	4	,,
Lark	4	to	4.30	,,
Black-headed titmouse	4.30	to	5	,,
Sparrow	5	to	5.30	,,

CHARACTERISTIC FLOWERS OF THE MONTHS.

January.

Christmas rose (*Helleborus*).
Winter aconite (*Eranthis hyemalis*).
Laurustinus (*Viburnum tinus*).

In the green-house, camellias are in full bloom.
In the stove, Strelitzia.
In the pits, forced hyacinths, tulips, and cinerarias.

February.

Snowdrop (*Galanthus nivalis*).
Crocus (*Crocus vernus*).
Hazel (*Corylus Avellana*).

March.

Primrose (*Primula vulgaris*).
Wood anemone (*Anemone nemorosa*).
Speedwell (*Veronica agrestis*).
Coltsfoot (*Tussilago Farfara*).

April.

Fruit blossoms generally, as

Apple (*Pyrus Malus*).
Pear (*Pyrus communis*).
Cherry (*Cerasus communis*).
Plum (*Prunus domestica*).
Apricot (*Prunus Armeniaca*).
Peach (*Amygdalus Persica*).
Almond (*Amygdalus communis*).

Narcissus (*Poet's and others*).
Daffodil (*Narcissus Pseudo-Narcissus*).
Cowslip (*Primula veris*).
Bluebell (*Hyacinthus nutans*).
Sweet violet (*Viola odorata*).
Elder (*Sambucus nigra*).

May.

Lady's-smock (*Cardamine pratensis*).
Strawberry (*Fragaria vesca*).
Hawthorn (*Cratægus Oxyacantha*).
Honeysuckle (*Lonicera Periclymenum*).
Buttercups (*Ranunculus*).
Daisy (*Bellis perennis*).
Lilac (*Syringa vulgaris*).
Laburnum (*Cytisus Laburnum*).
Crown Imperial (*Fritillaria imperialis*).

June.

Dog rose (*Rosa canina*).
Pimpernel (*Anagallis arvensis*).
Borage (*Borago officinalis*).
Horse chestnut (*Æsculus Hippocastanum*).
Columbine (*Aquilegia vulgaris*).
Lily-of-the-valley (*Convallaria majalis*.)
Guelder rose (*Viburnum Opulus*).

July.

White lily (*Lilium candidum*).
Harebell (*Campanula rotundifolia*).
Sunflower (*Helianthus annuus*).
Vine (*Vitis vinifera*).
Field Scabious (*Scabiosa arvensis*).
Many Thistles.
Corn poppy (*Papaver Rhœas*).
Common mallow (*Malva sylvestris*).
St. John's-wort (*Hypericum perforatum*).
Jasmine (*Jasminum officinale*).

August.

Toadflax (*Linaria vulgaris*).
Golden rod (*Solidago Virgaurea*).
China aster (*Aster Chinensis*).

Dodder (*Cuscuta Europæa*).
Teasel (*Dipsacus sylvestris*).
Chamomile (*Anthemis nobilis*).

SEPTEMBER.

Honeysuckle, flowers a second time (see May).
Ivy (*Hedera Helix*).
Ladies' tresses (*Ophrys spiralis*).

OCTOBER.

Strawberry-tree (*Arbutus Unedo*).
Chinese holly-oak (*Alcea rosea*).

Other characters of the months are given in the following table :—

JANUARY.

Berries on trees, as on Holly and Rose.
Trees without leaves. Buds on branches small.

FEBRUARY.

Buds on trees rather larger than last month.
Peas (*Pisum sativum*) sown. These are also sown in March.
Berries on bushes, few.

MARCH.

Oats (*Avena sativa*) sown. These are also sown in April.
Barley (*Hordeum sativum*) sown. Barley is sown till May.
Common elder (*Sambucus nigra*) comes into leaf in early seasons.
Gooseberry (*Ribes grossularia*) comes into leaf and bud in early seasons.
Ivy (*Hedera Helix*) berries ripe.
See February.

APRIL.

Elm (*Ulmus campestris*) comes into leaf if the season is early. This also applies to the two following trees :—
Beech (*Fagus sylvatica*), leafing.
Larch-tree (*Abies Larix*).
See March.

MAY.

This month particularly characterizes summer.

Walnut (*Juglans regia*) leafing.
White-beam-tree (*Pyrus Aria*) leafing.
Mulberry tree (*Morus nigra*).

In the case of trees leafing much is dependent upon the season. If it be late, all the trees here noted will develop their leaves a month or more later.

Summer or herbaceous stems shooting from the ground.

See April.

The flowers in this month are generally pale in colour; perhaps yellow prevails.

JUNE.

Buckthorn (*Rhamnus catharticus*) leafing.

There is a redundance of blossom in this month: a greater number of plants expand their blossoms in this month than in any other.

Trees in full leaf.
All foliage fresh, delicate, and full in colour.
Vegetable growth at its maximum.
Hay-making. (This goes into July.)
Puff-balls (*Lycoperdon bovista*) appear.
Truffles begin to be found.

JULY.

This month more particularly characterizes summer.

Whortleberries ripe (*Vaccinium uliginosum*).
Peas (*Pisum sativum*) cut. This may take place next month.
Apricots (*Prunus Armeniaca*) ripe. If the season is late, these are not ripe till August.
Rye harvest begins. This may be next month.
Wheat harvest begins in the earliest districts.
Flowers generally of bright colours.
Wild cherries ripe.

AUGUST.

Oats (*Avena sativa*) cut.
Barley (*Hordeum sativum*) cut.
Hop-picking begins.

These harvests are a month later in backward seasons.

Beech begins to turn yellow in early seasons.
Flowers few. Of bright colours.
Seeds of annuals ripe.

See July.

SEPTEMBER.

Apples, and fruits generally, ripe.
Grape harvest in early seasons.

Beans (*Vicia faba*) cut in early seasons.
See August.
Flowers almost gone; trees in autumn colours; seeds ripe.

OCTOBER—*Autumn.*

Wheat sown.
Fall of leaf. This may take place in November in late seasons.
Flowers gone; later seeds ripe.
Leaves of rich colours, if not shed.

NOVEMBER.

Trees without foliage.
Earlier berries ripe, as
Mountain ash.
Hawthorn.
Rose hips.

DECEMBER—*Winter.*

Later berries deepening in colour.

The following is a list of such plants as characterize the various kinds of soil:—

LAND PLANTS.

GEOGNOSTIC NATURE OF SOIL.

CHALK PLANTS.

Lady's-slipper (*Cypripedium Calceolus*).
Bee ophrys (*Ophrys apifera*).
Military orchis (*Orchis militaris*).
Pyramidal orchis (*O. pyramidalis*).
Teucrium montanum.
Blue moor-grass (*Sesleria cærulea*).
Box (*Buxus sempervirens*).

GRAVEL PLANTS.

Ordinary Gravel.
Alpine crowfoot (*Ranunculus alpestris*).
Ranunculus glacialis.
Baccharis phylicæformis.
„ *genistelloides.*
„ *sagittalis.*
„ *quadrangularis.*
Calycera ventosa.

- Land plants
  - Geognostic nature of soil, p. 198
    - Chalk, p. 198
    - Gravel
      - Ordinary, p. 198
      - Pit, p. 200
      - River, p. 200
    - Rock, p. 200
    - Sand
      - Sandy soil, p. 200
      - Drifting sands, p. 200
      - Sandy loam on clay subsoil, p. 201
      - Sandy loam on porous subsoil, p. 201
    - Clay
      - Good clay, p. 201
      - Thin clay, p. 201
      - Strong clayey loam on porous subsoil, p. 201
      - Thin clayey loam on porous subsoil, p. 201
    - Alluvial deposits, p. 202
    - Steppe, p. 202
  - Uncultivated soil, p. 202
    - Field, p. 202
    - Meadow, p. 203
    - Pasture
      - Low, p. 203.
      - High
      - Mountain
        - dry, p. 203
        - wet, p. 203
    - Heath
      - wet, p. 204
      - dry, p. 204
    - Moor
      - wet, p. 204
      - dry, p. 204
    - Marsh, p. 204
    - Bog, p. 200
    - Bush, p. 200
    - Forest, p. 200
    - Alpine, p. 200
    - Desert
      - Asiatic, p. 205
      - North African, p. 206
      - South African, p. 206
      - South American, p. 206
  - Cultivated soil, p. 207
    - Fallow, p. 207
    - Garden weeds, p. 207
    - Fence, p. 207
    - Hedge, p. 207
    - Field
      - Clay
        - Good clay, p. 208
        - Thin clay, p. 208
        - Clay on porous subsoil, p. 208
      - Sand
        - Sandy soil, p. 208
        - Sandy loam on clay subsoil, p. 208
        - Sandy loam on porous subsoil, p. 209
      - Alluvial deposits, p. 209
      - Reclaimed marsh, p. 209
      - Reclaimed heath, p. 209
  - Artificial productions, p. 209
    - Wall, p. 209
    - Roof, p. 210
    - Plank, p. 210
    - Rubbish, p. 210
- Earth and Water
  - Amphibious, p. 210
  - Inundated, p. 211
  - Sea-shore
    - Ordinary, p. 211
    - Marsh, p. 211
    - Gravel, p. 211
- Water plants
  - Salt, spring, p. 211
  - Fresh
    - Floating, p. 212
    - Aquatic
      - Lake, p. 212
      - Ditch, p. 212
      - River, p. 212
      - Spring, fresh, p. 212
    - Submerged, p. 213
  - Fresh and salt, p. 213

Sida pedicularifolia.
„ *borussica.*

Pit Gravel.

Knot grass (*Polygonum aviculare*).
Sheep's-sorrel (*Rumex Acetosella*).
Common bent-grass (*Agrostis vulgaris*).
Silvery hair-grass (*Aira caryophyllea*).
Hard fescue-grass (*Festuca duriuscula*).
Thyme-leaved sandwort (*Arenaria serpyllifolia*).
Wall hawkweed (*Hieracium murorum*).
Papaver dubium.
Climbing buckwheat (*Polygonum Convolvulus*).
Upright goosefoot (*Chenopodium urbicum*).
Perennial rye-grass (*Lolium perenne*).
Soft brome-grass (*Bromus mollis*).

River Gravel.

Brook saxifrage (*Saxifraga rivularis*).
Toad-rush (*Juncus bufonius*).
Sharp-flowered rush (*J. acutiflorus*).
Plantain shoreweed (*Littorella lacustris*).

Rock Plants.

Rock sedum (*Sedum rupestre*).
White sedum (*S. album*).
Welsh sedum (*S. Forsterianum.*)
Thlaspi saxatile.
Common thrift (*Statice Armeria*).
English scurvy-grass (*Cochlearia Anglica*).
Danish scurvy-grass (*C. Danica.*)
Mountain rose-root (*Rhodiola rosea*).

Sand Plants.

Sandy Soil.

Bird's-foot trefoil (*Lotus corniculatus.*)
Harebell (*Campanula rotundifolia.*)
Eyebright (*Euphrasia officinalis*).
Sweet-scented vernal grass (*Anthoxanthum odoratum*).

Drifting Sands.

Sea-bent (*Arundo arenaria*).
Sand-wheat grass (*Triticum junceum*).
Hard fescue grass (*Festuca duriuscula*).
Sand carex (*Carex arenaria*).
Hollow bedstraw (*Galium verum*).

Smooth rupture-wort (*Herniaria glabra*).
Sea mat-weed (*Arundo arenaria*).
Stonecrop (*Sedum acre*).
Sea lyme-grass (*Elymus arenarius*).
Ammophila arenaria.
Creeping willow (*Salix repens*),
Calamagrostis arenaria.
Buck's-horn plantain (*Plantago Coronopus*).

Sandy Loam on Clay Subsoil.

Common rush (*Juncus effusus*).
Sneezewort (*Achillea Ptarmica*).
Wild tansy (*Potentilla anserina*).
Mugwort (*Artemisia vulgaris*).

Sandy Loam on porous Subsoil.

Common Broom (*Genista scoparius*).
Black knapweed (*Centaurea nigra*).
Common ragweed (*Senecio jacobea*).
Hollow bedstraw (*Galium verum*).

Clay Plants.

Good Clay.

Queen of the meadows (*Spiræa Ulmaria*).
Wild sylvestris (*Angelica sylvestris*).
Great spearwort (*Ranunculus Lingua*).
Common sorrel (*Rumex Acetosa*).

Thin Clay.

Bitter crowfoot (*Ranunculus acris*).
Tufted hair-grass (*Aira cæspitosa*).
Corn horsetail (*Equisetum arvense*).
Marsh woundwort (*Stachys palustris*).

Strong clayey Loam on porous Subsoil.

Bladder campion (*Silene inflata*).
Toadflax (*Antirrhinum linaria*).
Red shank (*Polygonum amphibium*).
Great napweed (*Centaurea scabiosa*).
Rough cock's-foot grass (*Dactylis glomerata*).

Thin clayey Loam on porous Subsoil.

Common rest-harrow (*Ononis arvensis*).
Harefoot trefoil (*Trifolium arvense*).
Hop trefoil (*T. procumbens*).

Alluvial Deposits.

Common reed (*Arundo Phragmites*).
Round-headed rush (*Juncus conglomeratus*).
White bent-grass (*Agrostis alba*).
Reed meadow-grass (*Poa aquatica*).
Floating meadow-grass (*P. fluitans*).

Steppe Plants.

Dactylis repens.
Gynodon dactylon.
Zygophyllum album.
Cressa cretica.
Uniola maritima.
Spartina glabra.
Gerardia maritima.
Aster subulatus.
Poa thalassica.
Salsola corticosa.
„ *glomerata.*
„ *prostrata.*
Statice tartarica.
Glycirrhiza hirsuta.
„ *lævis.*
Atriplex lacinata.
Ceratocarpus arenarius.
Polycnemum.
Chenopodium.
Frankenia.
Tamarix.
Salicornia.
Chorispora sibirica.
Halocnemum.

PLANTS OF UNCULTIVATED SOILS.

Field Plants.

Common whitlow-grass (*Draba verna*).
Blunt-fingered speedwell (*Veronica triphyllos*).
Ivy-leaved speedwell (*V. hederifolia*).
Common viper's bugloss (*Echium vulgare*).
Common daisy (*Bellis perennis*).
Dandelion (*Taraxacum Dens-leonis*).
Buttercups.

Meadow Plants.

Clustered bell-flower (*Campanula glomerata*).
Spreading bell-flower (*Campanula patula*).
Myosotis scorpioides.
Trifolium pratense.
„ *fragiferum.*
„ *repens.*

Pasture Plants.

Low Pastures.

Daisy (*Bellis perennis*).
Common burnet-saxifrage (*Pimpinella Saxifraga*).
Ranunculus repens.
„ *bulbosus.*

High Pastures.

Common ling (*Calluna vulgaris*).
Hairy milk-vetch (*Astragalus uralensis*).
Mountain avens (*Dryas octopetala*).
Reticulated willow (*Salix reticulata*).
Mountain cudweed (*Gnaphalium alpinum*).
Cloudberry (*Rubus Chamæmorus*).
Common bear-berry (*Arbutus uva-ursi*).

Mountain Pastures (Dry).

Common ling (*Calluna vulgaris*).
Crow-berry (*Empetrum nigrum*).
Cross-leaved heath (*Erica Tetralix*).
Common club-moss (*Lycopodium clavatum*).
Savin-leaved club-moss (*L. alpinum*).
Stool-bent (*Juncus squarrosus*).
Paddock-pipe (*Equisetum palustre*).
Deer-hair (*Scirpus cæspitosus*).
Yellow-grass (*Narthecium ossifragum*).
Purple melic-grass (*Melica cærulea*).
Blue moor-grass (*Sesleria cærulea*).
Mat-grass (*Nardus stricta*).

Mountain Pastures (Wet).

Soft rush (*Juncus effusus*).
Soft meadow-grass (*Holcus mollis*).
Rup (*Juncus cæspitosa*).
Sprat (*J. acutiflorus*).
Pry (*Carex panicea*).
Devil's-bit scabious (*Scabiosa succisa*).

Heath Plants (Dry).

Common heath (*Erica vulgaris*).
Cross-leaved heath (*E. Tetralix*).
Fine-leaved heath (*E. cinerea*).
Common ling (*Calluna vulgaris*).
Dog's bent-grass (*Agrostis canina*).
Juniper (*Juniperus vulgaris*)
Ledum palustre.
Andromeda polifolia.
Parnassia palustris.
Sphagnum.
Polytrichum.

Heath (Wet).

Cotton-grass (*Eriophorum polystachion*).
Orange berry (*Vaccinium Oxycoccus*).

Moor Plants (Dry).

Petty whin (*Genista anglica*).
Mat-grass (*Nardus stricta*).
Yellow mountain-violet (*Viola lutea*).
Common tormentil (*Tormentilla officinalis*).
Mountain cudweed (*Gnaphalium dioicum*).

Moor (Wet).

Dwarf silky-willow (*Salix repens*).
Common butterwort (*Pinguicula vulgaris*).
Round-fruited carex (*Carex pilulifera*).
Moss-rush (*Juncus squarrosus*).
Scaly-headed club-rush (*Scirpus cæspitosus*).
Grass of Parnassus (*Parnassia palustris*).

Marsh Plants.

Ragged Robin (*Lychnis Flos Cuculi*).
Common buckbean (*Menyanthes trifoliata*).
Marsh marigold (*Caltha palustris*).
Brook-lime (*Veronica Beccabunga*).
Marsh cinquefoil (*Comarum palustre*).
Marsh bedstraw (*Galium uliginosum*).
Water featherfoil (*Hottonia palustris*).
Marsh fleawort (*Cineraria palustris*).
Nodding bur-marigold (*Bidens cernua*).
Cotton-grass (*Eriophorum angustifolium*).

Bog Plants.

Alpine butterwort (*Pinguicula alpina*).
Black bog-rush (*Schœnus nigricans*).
Downy-stalked cotton-grass (*Eriophorum pubescens*).
Bird's-eye primrose (*Primula farinosa*).

Bush Plants.

Common marjorum (*Origanum vulgare*).
Asarum europæum
Corydalis bulbosa.
Asclepias vincetoxicum.

Forest Plants.

Two-flowered linnæa (*Linnæa borealis*).
Pyrolæ.
Bilberry (*Vaccinium Myrtillus*).
Ophrys ovata.
Deadly nightshade (*Atropa belladonna*).
Water avens (*Geum rivale*).
European chickweed wintergreen (*Trientalis europæa*)
Common wood-sorrel (*Oxalis Acetosella*).
Orobanche.
Orchidæ.
Carex.

Alpine Plants.

Drooping saxifrage (*Saxifraga cernua*).
Alpine brook-saxifrage (*Saxifraga rivularis*).
Purple saxifrage (*Saxifraga oppositifolia*).
Whitlow grass (*Draba rupestris*).
Luzula arcuata.
Jointed pipewort (*Eriocaulon septangulare*).
Alsine rubella.
Pyramidal bugle (*Ajuga pyramidalis*).
Scottish primrose (*Primula Scotica*).
Least willow (*Salix herbacea*).
Alpine stitchwort (*Stellaria cerastoides*).

Desert Plants.

Asiatic Deserts.

Zizyphus Spina Christi.
Euphorbia antiquorum.

Pride of India (*Melia Azedarach*).
Calligonum polygonoides.
Anastatica hierochuntina.
Mimosa.
Tamarix.

North African.

Mesembryanthemum copticum.
„ *crystallinum.*
„ *biflorum.*
Acacia vera.
Sidra nebach.
Zizyphus Lotus.
Tamarix gallica.
„ *africana.*
„ *orientalis.*

South African.

Euphorbia meloformis.
„ *mammillaris.*
„ *biglandulosa.*
„ *trigona.*
„ *clava.*
„ *lactea.*
„ *neriifolia.*
Aloe ichotoma.
Stapeliæ.
Syngenesia.
Mesembryanthemum.
Sesuvium.
Crassula.
Sempervivum.
Sedum.
Cotyledon.
Bryophyllum.
Encephalartos.

South American.

Caven bush (*Acacia caven*).
Loranthus phyllus.
Cactus senilis.
„ *ficus-indica.*
Furcræa gigantea.
„ *longæva.*
Dwarf verbenas.

Opuntia tunas.
Cerei.
Cacti.
Melocactus.
Echinocactus.
Mammillaria.
Azorella.
Bolax.
Fragosa.
Lycopodia.
Yucca.
Agaves.

CULTIVATED SOILS.

FALLOW PLANTS.

Sorrel (*Rumex Acetosella*).
Field convolvulus (*Convolvulus arvensis*).
Androsace septentrionalis.
Echium vulgare.
Artemisia campestris.

GARDEN WEEDS.

Nettle (*Urtica urens*).
Alsine media.
White dead-nettle (*Lamium album*).
Hen-bit dead-nettle (*L. amplexicaule*).
White goosefoot (*Chenopodium album*).
Upright goosefoot (*C. urbicum.*)
Common groundsel (*Senecio vulgaris*).
Mouse-ear chickweed (*Cerastium arvense*).
Herb-mercury (*Mercurialis perennis*).

FENCE PLANTS.

Wild succory (*Cichorium Intybus*).
Common tansy (*Tanacetum vulgare*).
Mugwort (*Artemisia vulgaris*).
Yellow bedstraw (*Galium verum*).

HEDGE PLANTS.

Stinging-nettle (*Urtica dioica*).
White dead-nettle (*Lamium album*).
Borage (*Borago officinalis*).
Bryony (*Bryonia dioica*).
Broad-leaved bur-weed (*Xanthium strumarium*).
Thorn-apple (*Datura Stramonium*).

Violet (*Viola tricolor*).
Black bryony (*Tamus communis*).
Wood-sorrel (*Oxalis Acetosella*).
Blackberry (*Rubus fruticosus*).
Traveller's-joy (*Clematis Vitalba*).
Honeysuckle (*Lonicera Caprifolium*).
Everlasting-pea (*Lathyrus sylvestris*).
Tuberous moschatel (*Adoxa Moschatellina*).

Field Plants.

Clay Soil.

Good clay.

Common broad-leaved dock (*Rumex obtusifolius*).
Groundsel (*Senecio vulgaris*).
Nipplewort (*Lapsana communis*).
Corn-cockle (*Agrostemma Githago*).
Wild chamomile (*Matricaria Chamomilla*).

Thin clay.

Common coltsfoot (*Tussilago Farfara*).
Wild mustard (*Sinapis arvensis*).
Knot-grass (*Polygonum aviculare*).

Clay on porous Subsoil.

Scarlet pimpernel (*Anagallis arvensis*).
Ivy-leaved speedwell (*Veronica hederifolia*).
Black mustard (*Sinapis nigra*).
Hairy tare (*Ervum hirsutum.*)

Sand Plants.

Sandy Soil.

Common spurrey (*Spergula arvensis*).
Purple dead-nettle (*Lamium purpureum*).
Common fumitory (*Fumaria officinalis*).
Shepherd-purse (*Thlaspi Bursa-pastoris*).
Common knawel (*Scleranthus annuum*).
Common couch-grass (*Triticum repens*).
Common cudweed (*Gnaphalium germanicum*).

Sandy Loam on Clay Subsoil.

Charlock (*Raphanus Raphanistrum*).
Sheep-sorrel (*Rumex Acetosella*).

Corn-marigold (*Chrysanthemum segetum*).
Toad rush (*Juncus bufonius*).

Sandy Loam on porous Subsoil.

Common corn mint (*Mentha arvensis*).
Blue-wort (*Centaurea cyanus*).
Corn madder (*Sherardia arvensis*).
Corn gromwell (*Lithospermum arvense*).
Parsley-pest (*Alchemilla arvensis*).
Tall oat-grass (*Avena elata*).
Corn-thistle (*Cnicus arvensis*).

Alluvial Deposits.

Common reed (*Arundo Phragmites*).
Corn thistle (*Cnicus arvensis*).

Reclaimed Marsh Land.

Common colt's-foot (*Tussilago Farfara*).
Common butter-bur (*Petasites hybrida*).
Goosegrass (*Galium Aparine*).

Reclaimed Heath Land.

Soft brome-grass (*Bromus mollis*).
Field scorpion-grass (*Myosotis arvensis*).
Wild oats (*Avena fatua*).
Goosegrass (*Galium Aparine*).

ARTIFICIAL PRODUCTIONS.

Wall Plants.

Wall-germander (*Teucrium Chamædrys*).
Lecanora muralis.
Dicranum murale.
Wall-rue spleenwort (*Asplenium ruta-muraria*).
Rock spleenwort (*A. fontanum*).
Stonecrop (*Sedum acre*).
Live-long (*Sedum telephium*).
Ivy (*Hedera Helix*).
Rue-leaved saxifrage (*Saxifraga tridactylites*).
Wild wall-flower (*Cheiranthus fruticulosus*).
Linaria Cymbalaria.
Speedwell-leaved whitlow grass (*Draba muralis*).

Wall hawkweed (*Hieracium murorum*).
Jagged chickweed (*Holosteum umbellatum*).
Grimmia pulvinata.
Tortula muralis.

Roof Plants.

Houseleek (*Sempervivum tectorum*).
Mosses.
Lichens.

Board Plants.

Parmelia parietina.
Lecanora muralis.
Ramalina fraxinea.

Rubbish Plants.

Nettle (*Urtica dioica*).
Common dock (*Rumex obtusifolius*).
Mugwort (*Artemisia vulgaris*).
Annual poa (*Poa annua*).
Field poa (*P. pratensis*).
Common tansy (*Tanacetum vulgare*).
Round-leaved goosefoot (*Chenopodium polyspermum*).
Upright goosefoot (*C. urbicum*).
Red goosefoot (*C. rubrum*).
Stinking groundsel (*Senecio viscosus*).
Borage (*Borago officinalis*).
Burweed (*Xanthium strumarium*).
Common henbane (*Hyoscyamus niger*).

EARTH AND WATER PLANTS.

Amphibious Plants.

Lady's smock (*Cardamine pratensis*).
Water cress (*Nasturtium officinale*).
Amphibious yellow cress (*N. amphibium*).
Creeping yellow cress (*N. sylvestre*).
Great water-dock (*Rumex Hydrolapathum*).
Tall red mint (*Mentha rubra*).
Spear mint (*M. viridis*).
Horse mint (*M. sylvestris*).
Amphibious polygonum (*Polygonum amphibium*).

Inundated Plants.

Common mudwort (*Limosella aquatica*).
Water purslane (*Peplis Portula*).
Toad rush (*Juncus bufonius*).
Marsh marigold (*Caltha palustris*).
Water crowfoot (*Ranunculus scleratus*).

Sea-shore Plants.

Ordinary.

Sea-side campion (*Silene maritima*).
Sea plantain (*Plantago maritima*).
Black saltwort (*Glaux maritima*).
Sea lungwort (*Pulmonaria maritima*).
Sea holly (*Eryngium maritimum*).
Glass-wort (*Salsoli Kali*).
Sea-reed (*Arundo arenaria*).
Sea lyme-grass (*Elymus arenarius*).
Sea-side sedge (*Carex arenaria*).
Lythrum maritimum.
Common brook-reed (*Samolus Valerandi*).
Mangrove (*Rhizophora Mangle*).
Avicennia nitida.
„ *tomentosa.*

Marsh.

Sea arrow-grass (*Triglochin maritimum*).
Sea marsh-grass (*Poa procumbens*).
Pale carex (*Carex pallescens*).
Great common carex (*C. riparia*).

Gravel.

Sea rocket (*Cakile maritima*).
Sea-side goosefoot (*Chenopodium maritimum*).
Frosted sea orache (*Atriplex laciniata*).
Sea-side campion (*Silene maritima*).

WATER PLANTS.

Salt-water Plants.

Sea ruppia (*Ruppia maritima*).
Fucus pyrifera.
Sargassum vulgare.
Algæ.
Zosteræ.
Fuci.

Salt-Water Springs.

Glaux maritima.
Salsola.
Salicornia.
Anabasis.
Charæ.

Fresh-Water Plants.

Floating Plants.

Floating alisma (*Alisma natans*).
Common frogbit (*Hydrocharis Morsus-ranæ*).
Ivy-leaved duckweed (*Lemna trisulca*).
Lesser duckweed (L. minor).
Pistia stratiotes.
Confervæ.
Oscillatorieæ.

Aquatic Plants.

Lake Plants.

White water-lily (*Nymphœa alba*).
Yellow water-lily (*Nuphar luteum*).
Bull-rush (*Scirpus lacustris*).
Scirpus palustris.
Common reed (*Arundo Phragmites*).
Water soldier (*Stratoites aloides*).
Butomus umbellatus.
Pontederia Lotus.
Water crowfoot (*Ranunculus aquatilis*).

Ditch Plants.

Common frogbit (*Hydrocharis Morsus-ranæ*).
Forget-me-not (*Myosotis palustris*).
Water speedwell (*Veronica Anagallis*).
Phellandrium aquaticum.
Water soldier (*Stratoites aloides*).
Common flowering rush (*Butomus umbellatus*).

River Plants.

Ranunculus fluviatilis.
Conferva rivularis.

Fresh-water Springs.

Water chickweed (*Montia fontana*).
Brooklime (*Veronica Beccabunga*).

SUBMERGED PLANTS.

Hornwort (*Ceratophyllum demersum*).
Chara.
Awlwort (*Subularia aquatica*).
Vallisneria spiralis.

FRESH AND SALT WATER PLANTS.

Conferva glomerata.
Slender pondweed (*Potamogeton pusillus*).
Fennel pondweed (*P. pectinatus*).
Common Zannichellia (*Zannichellia palustris*).
Diatomeæ.

Each of the following plants covers a large tract of country by the frequency of its repetition. The plants named in this list have social habits, as they live in groups.

SOCIAL PLANTS.

Sphagnum palustre.
Dicranum glaucum.
Arundo Phragmites.
Pinus sylvestris,
Polygonum aviculare.
Polygonum junceum
Poa annua
Vaccinium Myrtillus.
Juncus bufonius,
Myriophyllum spicatum.
Alnus glutinosa.
Betula alba.
Rhizophora Mangle.
Banksia speciosa.
Protea argentea.
Protea mellifera.
Bambusa arundinacea
Thymus cover.
Erica vulgaris.
Acacia caven.
Lycium gracile.
Cupressus callitris.
Cereus peruvianus.
Cocos nucifera.
Musa sapientum.
Musa paradisiaca.
Dracæna terminalis.
Pandanus odoratissimus.
Nipa frutescens.
Chamærops humilis.
Phœnix dactylifera.
Mauritia flexuosa.
Pteris aquilina.
Bæckia frutescens.
Tilia europea.
Fagus sylvatica.
Castanea vesca.
Quercus robur.
Ulmus campestris.
Opuntia tunas.
Rhododendron.
Araucaria excelsa.
Eucalyptus globulus.
Mimosæ.
Cinchona.
Escallonia.
Genista.
Ulex.
Baccharidæ.

The following lists give the characteristic plants of different countries :—

EUROPE.

France.

In the South (wild).
- Evergreen oaks.
- Myrtles.
- Wild olives.
- Cistus.
- Dwarf palm.
- Indian fig.
- Aloe.

(This district is particularly characterized by evergreen trees and shrubs.)

Middle and North (wild).
- Pinus maritima.
- Common heath (*Erica vulgaris*).

The pea blossoms and plants allied to our dandelion and thistles are the most common plants. In this district the trees shed their leaves.

In cultivation.
- Olive.
- Maize.
- Vine.
- Potato.
- Orchard trees.

Spain and Portugal.

The South.
- Palms.
- Date.
- *Chamœrops humilis* (a fan palm).
- Banana (in the south).
- Orange (flowers April and May).
- Lemon.
- Olive.
- Pomegranate.
- Fig.
- Mulberry.
- Cypress.
- Judas tree.
- Cactus (*Cactus Tuna*).
- Aloe (American).
- Vine.
- Meadow saffron.
- Autumnal snowdrop.
- Gum cistus.

Forest trees.
- *Quercus Ballota* (a kind of oak).
- „ *œgilopifolia* „
- „ *faginea* „
- „ *prasina* „
- „ *crenata* „
- „ *rotundifolia* „
- „ *humilis* „
- „ *suber* (Cork oak).
- Lime tree.
- Beech.
- Chestnut (Spanish).
- Yew.
- Scotch pine.

(The forest trees are chiefly evergreens.)

Shrubs and small trees.
- Bay tree.
- Myrtle.
- *Phillyrea media.*
- „ *angustifolia*
- *Juniperis Sabina.*

Celtis australis.
Pistacia.
Terebinth.
Lentiscus.
Rhamnus Alaternus.
Strawberry tree.
Jasmine.
Caper.

BALEARIC ISLES.

Cultivated.
Cotton.
Orange.
Olive.
Carob tree.
Fig.
Almond.
Date palm.
Cactus Opuntia.
Vine.
Some species of grain.

Wild.
Pinus halepensis.
Oak (common).
Clematis cirrhosa.
Hypericum balearicum.
The dwarf palm.
Cyclamen.
Polygala.
Anonis.
Anthyllis.

ITALY AND SICILY.

The South (wild).
Aloe (*A. americana*).
Prickly fig.
Orange.
Citron.
Olive.
Myrtle.
Laurel.
Carob tree.
Pomegranate.
Arbutus.
Tamarisk.
Manna ash.
Chestnut (edible).
Mulberry.
Poplar.
Stone pine (*Pinus Pinea*).
Lemon.
Caper.
Dwarf palm.
Pinus halepensis.

North (wild).
Common oak.
Beech.
Ash.
Horse chestnut.
Plum.
Cork oak (*Quercus suber*).
Quercus cerris.
Yew.
Larch.
Scotch fir.
Pinaster.
Arundo Donax (a large and important reed).

Cultivated.
Wheat.
Maize.
Millet.
Rice.
Vine.
Mulberry (red and white).
Sugar cane.
Date.
Custard apple.

Greece and the Grecian Archipelago.

Trees and Shrubs (wild).
- Olive.
- Myrtle.
- Orange.
- Lemon.
- Laurel.
- Oaks (*Quercus Ilex*).
- „ („ *coccifera*).
- „ („ *Ægilops*).
- Common oak.
- Oriental hornbeam.
- Beech.
- Manna ash.
- Scotch fir.
- Stone Pine.
- Prickly fig.
- Date.
- *Osyris alba.*
- Terebinth.
- Caper.
- Jasmine.
- *Phillyrœa media.*
- „ *angustifolia.*
- Strawberry tree.
- Pomegranate.
- Christ's thorn.
- Poplar.
- Aspen.
- Chaste tree.
- Oleander.
- Yew.
- Ash.
- Chestnut (edible).
- Apple.
- Pear.
- Cypress.
- Almond.

Herbaceous, or tender plants.
- *Narcissus Tazetta.*
- *Loranthus europus.*
- Tuberoses.
- Hyacinths.
- Irids.
- Ranunculus.
- Tulip.
- *Anemone hortensis.*
- *Poterium spinosum.*
- *Dianthus arboreus.*

Cultivated plants.
- Pistacia.
- Lentiscus.
- *Astragalus tragacantha.*
- „ *cretica.*
- „ *terebinthus.*
- Vine.
- Wheat.
- Barley.
- Maize.
- Cotton.

Turkey.

Cultivated.
- Cotton.
- Wheat.
- Mulberry.
- Maize.
- Much pasture land.

Switzerland.

Wild.
- Siberian pine (*Pinus Cembra*)
- Pinaster.
- Larch.
- Alder.
- Maple.
- Ash.
- Cytisus.

Broom.
Sorrel.
Elm.
Oak.
Spruce.
A number of pea-blossoms.

Cultivated.
Wheat.
Maize.
Potato.
Barley.
Rye.
Walnut.
Mulberry.
Chestnut (Spanish).
Hemp.
Beans.
Much pasture land.

GERMANY, *including Prussia, Austria, Hungary, and the German States.*

Wild.
Common heath (*Erica vulgaris*). (This is very common.)
Juniperis communis.
Ledrum palustre.
Andromeda polifolia.
Abietinæ.

The forests are formed of the ordinary trees, as birch, beech, &c., which shed their leaves.

Honeysuckle.
Ivy.
Rose.
Hop.
Rubus (the blackberries).
Guelder rose.
Trapa natans.
Chara vulgaris.
Salvia pratensis.
Thesium linophyllum.
Limosella aquatica.
Cucubalus tatarica.
(Common in forests.)

One-third of Germany is forest land.

Cultivated.
Rye.
Buckwheat.
Wheat.
Barley.
Oats.
Maize.
Peas.
Beans.
Lentils.
Cabbage.
Flax.
Hops.
Rape.
Vine.

HOLLAND AND BELGIUM.

Wild.
Common heath.
Fir.
Waterlily (white & yellow).
Flowering rush.

Cultivated.
Grasses.
Clover.
Madder.
Rape.

Mustard.
Hop.
Ordinary vegetables.
Wheat.
Rye.
Buckwheat.
Barley.
Oats.
Pulse.
Potatoes.
Hemp.
Flax.

Denmark.

Wild.
- Pines.
- Firs.
- Common heath.
- Juniper.
- *Ledum palustre.*
- *Andromeda polifolia.*
- Willows.
- Birch.
- Beech.
- Oak.
- Elm.

Cultivated.
- Wheat.
- Barley.
- Oats.
- Rye.
- Potato.
- Flax.
- Hemp.
- Tobacco.
- Much pasture land.

Sweden and Norway.

Wild.
- A great preponderance of forest land.
- Birch.
- Pine (*Pinus sylvestris*).
- Spruce.
- Poplar.
- Willow.
- Aspen.
- Juniper.
- Ash.
- Lime.
- Elm.
- Foxglove.
- *Hieracium aurantiacum.*
- *Sedum anglicum.*
- *Chrysosplenium oppositifolium.*
- *Hypericum pulchrum.*
- *Erica cinerea.*
- *Rosa spinosissima.*

Cultivated.
- Rye.
- Oats.
- Bigg.
- Barley.
- Pulse.
- Potato.

Russia (*in Europe*).

Southern part.

Wild.
- Common oak.
- Beech (red and white).
- Sea pine (*Pinus maritima*).
- Juniper.
- Yew.
- Quercus cerris.
- Beech.
- Elm.
- Poplar.

Linden tree.
Plane.
Maple.
Ash.
Apple.
Pear.
Manna ash.
Alder.
Spindle tree.
Water elder.
Wayfaring tree.
Lantana.
Rose.
Privet.
Cornel.
Vine.
Trefoil.
Clover.
Medick.
Lucerne.
Madder.
Goosegrass.
Woad.
Dyer's green.
Litmus.
Safflower.
Sumach.
Cotinus.

On the extensive Steppes.
Atriplex lacinata.
Astragali.
Glycirrhiza hirsuta.
" *lævis.*
Glycirrhiza echinuta.
Salsola prostrata.
Statice tartarica.
Lathyrus tuberosus.
Medicago sativa.
Vicia sylvatica.
Lotus corniculatus.
Serratula arvensis.
Inula brittanica.
Artemisia absinthium.
Anabasis aphylla.
Tamarix gallica.
Cynanchum acutum.
Senecio linifolius.

Cultivated.
Vine.
Wheat.
Rye.
Barley.
Maize.
Millet.
Chick peas.
Flax.
Tobacco (*Nicotiana paniculata*).
Onions.
Apples.
Pears.
Plums.
Figs.
Walnuts.
Mulberry.
Hazel-nut.

Northern part.

Wild.
Scotch fir (forming immense forests).
Birch.
Lime.
Beech.
Maple.
Elm.
Alder.
Willow.
Ash.
Mountain ash.
Hazel.
Willow.
Juniper.
Cranberry.

Bilberry.
Cultivated.
Rye.
Barley.
Oats.
Hemp.
Flax.
Potato.

Lapland.

Wild.
Spruce fir.
Scotch fir.
Birch.
Meadow trefoil.
Lily of the valley.
White water-lily.
Yellow „ „
Salix glauca.
„ *hastata.*
Reindeer moss (*Cenomyce rangiferina*).
Iceland moss (*Lichen islandicus*).
Cudbear.
Arbutus alpina.
Juncus trifidus.
Lycopodium alpinum.
Prunella vulgaris.
Viola bifolia.
Aspen.
Bird-cherry.
Mountain ash.
Diapensia lapponica.
Red currant.
Arctic raspberry.
Cultivated.
Barley.

Iceland and Faroe Islands.

Wild. No trees.
Willow.
Birch } shrubby.
Alder }
Juniper.
Grasses.
Erica vulgaris.
Chara.
Hippuris vulgaris.
Veronica Anagallis.
Arundo Phragmites.
Comarum palustre.
Limosella aquatica.
Serratula arvensis.
Shepherd's purse.
Draba verna.
Prunella officinalis.
Thymus serpyllum.
Lychnis Flos Cuculi.
Spergula arvensis.
Vaccinium Myrtillus.
„ *uliginosum.*
„ *Oxycoccus.*
Arbutus Uva-ursi.
Iceland moss.
Pisum maritimum.
Arundo arenaria.
Cultivated.
Barley.
Potato.
Turnip.
Cabbage.
Cauliflower.
Cummin.

ASIA.

Asia Minor and Palestine.

Wild.
- Sweet chestnut.
- Cherry.
- Fig.
- Olive.
- Mulberry.
- Rose.
- Horse chestnut.
- Lilac.
- Sweet jasmine.
- Melon.
- Cucumber.
- Willow.
- Date.
- Henna.
- Banana.
- Sugar-cane.
- Orange.
- Citron.
- Pistachio.
- Carob-tree.
- *Cordia mixa.*
- *Guilandina morinda.*
- Indian tamarind.
- *Melia Azedarach.*
- *Acacia nilotica.*
- ,, *farnesiana.*
- Apple of Sodom (*Solanum sodomæum*).
- Oak.
- Pine.
- Plane.
- Pear.
- Fir.
- Beech.
- Apple.
- Plum.
- Peach.
- Medlar.
- Quince.
- Walnut.
- Myrtle.
- Strawberry tree.
- Laurel.
- Pomegranate.
- Terebinth.
- Oleander.
- Vine.
- Sycamore fig (*Ficus Sycamorus*).
- Cedar of Lebanon.
- Christ's thorn (*Zizyphus Spina Christi*).
- *Amyris gileadensis*
- Rose of Jericho (*Anastatica hierochuntica*).

Cultivated.
- Wheat.
- Rye.
- Barley.
- Beans.
- Maize.
- Cotton.
- Sugar-cane.
- Indigo.
- Tobacco.
- Mulberry.

Arabia.

Wild.
- Fig.
- *Keura odorifera.*
- Pear.

Orange.
Lemon.
Date.
Apricots.
Peach.

Cultivated.

Coffee.
Wheat.
Durra (*Holcus Sorghum*).
Balsam (*Amyris opobalsamum*.
Myrrh (*Amyris Kataf*).
Aloes (*Aloe officinalis*).
Senna (*Cassia lanceolata*).
Tamarinds (*Tamarindus officinalis*).
Gourd.
Melon.
Cocoa-nut.
Pomegranate.
Almond.
Filbert.
Mangosteen.

PERSIA.

Wild.

Mimosa.
Christ's thorn (*Zizyphus Spina Christi*).
Water melons.
Tamarix.
Euphorbia antiquorum.
Date.
Pride of India (*Melia Azedarach*).
Banyan.
Dalbergia Sissoo.
Mango.
Tamarind.
Walnut.
Sycamore.
Oriental plane.
Guava.
Banana.
Fig.
Pistachio.
Mulberry.
Pomegranate.
Vine.
Quince.
Peach.
Apricot.
Almond.
Cherry.
Currants.
Lentisk.
Orange.
Cypress.

Cultivated.

Mulberry.
Cotton.
Indigo.
Sesame.
Holcus spicatus.
Rice.
Sorghum saccharatum.
Wheat.
Barley.
Maize.
Beet.
Carrot.

HINDOSTAN.

Wild.

Butea frondosa.
Careya.
Chestnut.
Hornbeam.
Cedrelas.

Croton.
Daphne gardneri.
„ *cannabinum.*
Dipteronitis grandiflora.
Spindle tree.
Ash (*Fraxinus floribunda*).
Gmelina arborea.
Walnut.
Teak.
Holly.
Juniper.
Laurel.
Privet.
Nutmeg.
Magnolia.
Olive.
Pinus Deodara.
Rhododendron arboreum.
Willow.
Pium.
Oaks (10 species).
Pyrus.
Bamboo.
Sandal tree.
Sacred bean.
Chamœrops Martiana.
Cocoa-nut.
Palmyra palm.
Caryota urens.
Betel-nut palm.
Phœnix farinifera.
„ *sylvestris.*
„ *humilis.*
Sago palm (*Sagus farinifera*).
Mango.
Orange.
Pomegranate.
Peach.
Banana.
Pancratium.
Crinum.
Orchids.

Cultivated.
Cotton.
Sugar.
Indigo.
Pine apple.
Bamboo.
Rice.
Maize.
Opium poppies.
Coffee.
Betel pepper.
Areca-nut palm.
Holcus sorghum.
Panicum.
Paspalum.
Elansinæ.
Phaseolus.
Dolichas.
Gourds.
Tesamum.
Ginger.
Turmeric.
Crotolaria juncea.
Hibiscus cannabinus.
Tobacco.
Carrots.
Barley.
Millet.
Wheat.
Oats.
Beans.
Vetches.
Mustard.
Coriander.
Flax.
Saffron.

The following are plants of Ancient Hindostan :—

Nymphæa stellata.
Nelumbium.
Banana.
Acacia.
Date palm.
Pandanus.
Mango.
Sacred fig.
Bamboo.
Rice.
Barley.
Wheat.
Myrtle.
Fig.
Sugar-cane.
Hemp.
Flax.
Lentil.
Onion.
Garlic.
Leek.
Cotton.
Henna.

China.

Wild.
- Palms.
- Laurel.
- Caper.
- Tea.
- Mallows.
- Bamboo.
- Camellias.
- Orange.
- Lemon.
- Magnolias.
- Terebinth.
- Myrtles.
- Aucuba.
- Oak.
- Hornbeam.
- Poplar.
- Oleander.
- Sacred Bean of India.
- *Pinus chinensis.*
- *Bæckia frutescens.*

Cultivated.
- Tea.
- Banana.
- Guava.
- Orange.
- Papaw.
- Cocoa.
- Litche.
- Sugar-cane.
- Peach.
- Apricot.
- Vine.
- Pomegranate.
- Rice.
- Mulberry (black and white).
- Walnut.
- Fig.
- Lemon.
- *Olea fragrans.*

Japan.

Wild.
- Tea.
- Camellias.
- *Lycium barbatum.*
- Lime.
- Scotch fir.
- *Pinus cembra.*
- „ *strobus.*
- Spruce.
- Larch.
- *Aucuba japonica.*
- *Salisburia adiantoides.*

Sophora japonica.
Pyrus japonica.
Mespilus japonica.
Hydrangea hortensis.
Yew.
Juniper.
Salix japonica.
Eleagnus umbellata.
Orange.
Camphor tree.
Bamboo.
Olea fragrans.
Cycas revoluta.
Chestnut.
Birch.
Alder.
Ash.
Mulberry (black and white)
Salix integra.
Betula japonica.
Quercus glabra.
,, *glauca.*
,, *acuta.*
,, *cuspidata.*
Quercus serrata.
Cupressus japonica.

Cultivated.
Rice.
Daid-su (a kind of bean).
Wheat.
Barley.
Turnips.
Mulberry.
Tea.
Fir.
Cypress.
Fig.
Chestnut.
Bamboo.
Convolvulus.
Dioscorea.
Polygonum fagopyrum.
Pomegranate.
Nymphœa nelumbo.
Arum esculentum.
Carrot.
Cycas revoluta.
Sesamum orientale.

Cashmere.

Wild.
Pine.
Fir.
Willows.
Poplar.

Cultivated.
Orchard trees.
Mulberry.
Rice.
Wheat.
Barley.

Thibet.

Wild.
Spruce.
Fir.
Juniper.
Oak.
Hazel.
Alder.
Willow.
Birch.
Elder.
Horse chestnut.
Ash.
Gooseberry.
Raspberry.
Rhododendrons.
Vaccinium.

Erythrina monosperma.
Bombax heptaphyllum.
Chestnut.
Pine.
Laurel.
Michelia.
Gordonia.

Cultivated.
Mango.
Orange.
Pomegranate.
Peach.
Apple.
Pear.
Walnut.
Banana.
Bamboo.
Pine-apple.
Sugar-cane.
Rice.
Opium poppy.
Mulberry.

SIBERIA.

Western Siberia with Eastern-side of the Altai Mountains.

Wild.

Forest trees.
Birch.
Aspen.
Poplar.
Larch.
Pinus Cembra.
Firs.
Pines.
Spruce.
Pinus sibirica.
„ *Abies.*

Mountain plants.
Gentiana altaica.
Veronica densiflora.
Primula nivalis.
Pedicularis resupinata.
Gentiana acaulis.
Cortusa Mathiola.
Cardamine macrophylla.
Saxifraga Geum.

Steppe plants.
Adonis vernalis.
Anemone patens.
Artemisia.
Allium.
Gypsophilla.
Statice.
Ceratocarpus arenarius.
Diotis ceratoides.
Polycnemum.
Atriplex.
Chenopodium.
Frankenia.
Tamarix.
Salicornia.
Halocnemum.
Amaryllis tatarica.
Rindera tetraspis.
Nepeta sibirica.

Central Siberia.

Wild.
Pinus Cembra.
„ *larix.*
Pinus Abies.
Plantanus orientalis.
Populus alba.

Populus balsamica.
Betula alnus.
  „ *nana.*
  „ *fruticosa.*
Rhododendron tauricum.
  „ *chrysanthum.*
Stachys palustris.
  „ *sylvatica.*
Scutellaria galericulata.
Schwertia perennis.
Sanguisorba officinalis.
Tanacetum vulgare.
Trientalis europæa.
Valeriana officinalis.
Vaccinium vitis-idæa.
  „ *uliginosum.*
Anemone narcissiflora.
  „ *sylvestris.*
Atragene alpina.
Andromeda polifolia.
Antirrhinum linaria.
Arbutus uva-ursi.
Euphrasia officinalis.
Potentilla anserina.
Pyrola.
Galium boreale.
Sedum palustre.
Lysimachia thrysiflora.

Eastern Siberia.

Wild.
Birch.
Pinus Cembra.
Sorbus aucuparia.
Alnus incana.
Willows.
Spiræa kamtschatica.
Allium ursinum.
Maianthemum canadense.
Uvularia amplexifolia.
Iris sibirica.
Caprifolium.
Rose.
Atragene alpina.
Rhododendron kamtschaticum.
Empetrum nigrum.
Trientalis europæa.
Linnæa borealis.
Cornus suecica.
Urtica dioica.
Lupinus nootkensis.
Mimulus luteus.
  „ *guttatus.*
Epilobium angustifolium.
  „ *latifolium.*
  „ *luteum.*
Rubus spectabilis.
Angelica.
Heracleum.
Carices.
Orchids.
Ferns.
Potamogeton.
Ranunculus aquatilis.
Sparganium.
Hippuris
Sorrel.
Polygonum.
Aconite.
Monkshood.
Meadow rue.
Geranium pratense.
Comarum palustre.
Oxycoccus.
Arbutus alpina.
Azalea procumbens.

East Indian Archipelago and Ceylon.

Wild.
Rafflesia Arnoldii.
Pitcher plant (*Nepenthes distillatoria*).

Cinnamon.
Cocoa-nut
Caryota.
Date palm.
Banana.
Mango.
Jack.
Malay apple.
Mangosteen.
Psidia.
Oranges.
Citron.
Cashew-nut.
Guava.
Averrhoas.
Amoma.
Hedysarum.
Galega.
Hibiscus.
Justicia.
Cleome.
Impatiens.
Myrtle.
Amomum.
Ricinus.
Ipomæa.
Dioscorea.
Basella.
Aristolochia.
Ophioglossum.
Phaseolus.
Memordica.
Casuarina.
Bryonia.
Vine.
Cissus.
Pothos.
Loranthus.
Acrostichum.
Ginger.
Cardamom.
Galanga.
Costus.
Acorus.
Schœnanthus.
Calamus aromaticus.
Turmeric.
Pandanus.
Nipa palm (*Nipa frutescens*).
Mangrove.
Avicenniæ.
Fig trees.
Melaleuca leucadendron.
Eugenia.

Cultivated.
Cinnamon.
Camphor.
Clove.
Nutmeg.
Pepper.
Betel pepper.
Ginger.
Galangale.
Rice.
Eugenia.

AFRICA.

Barbary.

Wild on the mountains.
Quercus suber.
„ *Ilex.*
„ *obtecta.*
„ *pseudo-suber.*
Quercus coccifera.
„ *pseudo-coccifera.*
„ *Ballota.*
Aleppo pine (*Pinus halepensis*.

Fresnella Fontanesii.
Juniperus phœnicea.
„ *lycia.*
Cypress.
Wild on the plains.
Buckthorn.
Oleander.
Olive.
Walnut.
Acacia.
Cedar.
Date palm.
Rose.
Peaches.
Plums.
Almonds.
Oranges.
Pomegranate.
Figs.
Vine.
Pistachio.
Zizyphus Lotus.
Mulberry.
Citron.
Myrtle.
Jasmine.
Spina Christi.
Chaste tree.
Palmetto Palm (Chamærops humilis).
Castor-oil tree.
Cultivated.
Wheat.
Barley.
Olive.
Vine.
Date palm.
Mulberry (white).

EGYPT AND NUBIA.

Wild by the Nile.
Date palm.
Figs.
Willows.
Acacias.
Tamarisks.
Sycamore fig.
Oleander.
Christ's thorn.
Papyrus.
Sandy plains.
Tamarisk.
Caper.
Cassias.
Acacias.
Mimosa.
Salsola.
Traganum.
Calligonum.
Salicornia.
Mesembryanthemum.
Colutea.
Carduus.
Prickly fig.
Oases.
Date palm.
Doum palm.
Palm of Thebes (Cucifera thebiaca).
Acacia vera.
Orange.
Citron.
Banana.
Olive.
Pomegranate.
Peach.
Cultivated.
Date palm.
Sycamore fig.
Egyptian arum (*Arum colocacia*).
Gourd.

Cucumber.
Sassflower.
Onion.
Rice.
Wheat.
Barley.
Holcus dhourra.
Beans.
Maize.
Flax.
Sugar.
Cotton.

Plants of Ancient Egypt.

Sacred bean of India.
Papyrus.
Red-flowered lotus.
Onion.
Bean (*Vicia faba*).
Vine.
Arundo donax.
Water-melon.
Cucumber.
Garlic.
Radish.
Date palm.
Doum palm.
Flax.
Olive.
Pear.
Barley.
Lentil.
Almond.
Leek (*Allium porum*).
Rose.
Cummin (*Cumimum cymimum.*
Poplar (white and black).
Aspen.
Melilotus officinalis.
Rye.
Opium poppy (*Papaver somniferum*).
Sassflower.
Ash.
Oak.
Alder.
Cypress.
Thuja orientalis.
Viola odorata.
Mallow.
Squill (*Scilla maritima*).
Cabbage.
Plum.
Cotton.
Hemp.
Beet.
Mustard.
Rue.
Ocymum basilicum.
Black mulberry.
Fennel.
Madder.
Coriander
Carrot.
Asparagus.
Endive.
Elder.
Ficus carica.
Pomegranate.
Zizyphus Lotus.
Myrtle.
Pistachio-nut.
Calendula officinalis.
Plantago major.

Abyssinia.

Wild.
Papyrus.
Amyris Opobalsamum.
Juga Sassa.

Banana.
Kol-qual (*Euphorbia antiquorum*).
Gir-gir (*Andropogon afer*).
Kantuffa.
Gaguedi (*Protea abyssinica*).
Farek (*Bauhinia acuminata*).
Kuara (*Erythrina abyssinica*).
Teff (*Poa abyssinica*).
Cassia fistula.

Central Africa.

Wild.
- Passion-flowers.
- Aristolochiæ.
- Calamus.
- Balanophoræ.

Cultivated.
- Millet.
- Cotton.
- Onions.
- Rice.
- Indigo.
- Wheat.
- Ground nuts.

Western Africa.

Wild.
- *Adansonia digitata.*
- *Bombax pentandrum.*
- *Raphia vinifera.*
- *Pandanus Candelabrum.*
- *Sterculia acuminata.*
- Ordeal tree.
- *Anona senegalensis.*
- *Chrysobalanus Icaco.*
- Pine-apples.
- Oil palm.
- Wild guava (*Psidium pyriferum*).
- Orange.

Cultivated.
- Maize.
- Cassava (*Jatropha Manihot*).
- Pulse.
- Ground-nut.
- Limes.
- Plantain (*Musa sapientum*).
- Papaw (*Carica Papaya*).
- Pine-apple.
- Pumpkin (*Cucurbita Pepo*).
- Tamarind.
- Oil-palm (*Elais guineensis*)
- Tobacco.
- Capsicum.
- Negro peach (*Sarcocephalus esculentus*).
- Locust-tree (*Inga biglobosa*)
- Cashew-nut.
- Baobab, or monkey-bread.

Southern Africa.

Wild.
- Heaths.
- Proteas.
- Silver-tree (*Protea argentea*).
- *Protea repens.*
- „ *mellifera.*
- Fig-marigolds.
- Stapelia.

Diosma.
Aloes
Euphorbias.
Tamus elephantopus.
Irids
Orchids.
Strelitzia.
Palmiet (*Juncus serratus*).
Pelargonium.
Acacias.
Polygala.
Fabaceæ.
Acacia capensis.
Asters.
Gnaphalium.
Xeranthemum.
Zamiæ.
Ixiæ.
Amaryllidæ.
Elichrysum.
Eriocephalus.
Calendula.
Othonna.
Arctotis.
Corymbium.
Senecio.
Virgilia.
Aspalathus.
Lobelia.
Indigofera.
Agathosma.
Morea.
Phylica.
Gladiolus.
Halleria.
Acorus palmita.
Cluytia.
Barrosma.
Cycadeæ.

MADEIRA.

Wild.
Vaccinium padifolium.
Prickly fig.
Myrtle.
Chestnut.
Spartium.
Heaths.
Pteris aquilina.
Cichoraceæ.

Cultivated.
Vine.
Fig.
Arum.
Banana.
Pomegranate.
Mulberry.
Potato.
Onions.

CANARY ISLANDS.

Wild.
Euphorbias.
Mesembryanthemums.
Cacalia Kleinia.
Dragon's blood tree.
Date.
Banana.
Palmetto.
Indian fig.
Laurel.
Visnea.
Olive (*Olea excelsa*).
Arbutus callicarpa.
Sideroxylon.
Ivy (*Hedera canariensis*).
Ferns.
Golden campanula (*Campanula aurea*).
Fir (*Pinus canariensis*).
Juniperis cedro.

Spartium nubigenum.
Grasses.
Cypress.
Erica Arborea.
Agaves.
Sempervivum.
Aizoon.
Cotyledon.
Crassula.
Cordon (*Euphorbia canariensis*).
Euphorbia balsamifera.

Cultivated.
- Sugar cane.
- *Arum calocacia.*
- Olive.
- Vine.
- Wheat.
- Barley.
- Maize.
- Chestnut.

Madagascar.

Wild.
- Tanghii tree (*Tanghinia veneniflua*).
- *Hydrogeton fenestralis.*
- Palms.
- Bamboo.
- Orange.
- Citron.

Cultivated.
- Rice.
- Maniæ.
- Batata.
- Maize.
- Calabastes.
- Earth-nut (*Arachis hypogea*).
- Sugar-cane.
- Pine-apple.
- Bread-fruit tree.
- Vine.
- Cotton.
- Hemp.
- Potato.
- Bean.

Mauritius and Bourbon.

The flora of these islands is almost identical with that of the Indian Archipelago.

Cultivated.
- Cloves.
- Nutmeg.
- Sugar-cane.
- Coffee.
- Cotton.
- Indigo.

NORTH AMERICA.

Mexico.

Wild.
- Vanilla.
- Caoutchouc.
- *Styrax officinalis.*
- Hand plant.
- Dahlias.
- *Convolvulus Jalapa.*
- *Agave americana.*

Logwood tree.
Furcrœa.
Yuccæ.

Cultivated.

Rice.
Maize.
Coffee
Cocoa.
Sugar-cane.
Manioc.
Banana.
Maguey.
Cactus coccinellifer.
Anona cherimolia.
Wheat.
Cotton.
Agaves.
Opuntia tunas.

SUB-TROPICAL ZONE OF NORTH AMERICA.

(From Tropics to 34°.)

Wild.

Magnolia grandiflora.
„ *glauca.*
Calycanthus floridus.
Kalmia hirsuta.
„ *cuneata.*
Halesia tetraptera.
„ *diptera.*
Laurus catesbeyana.
„ *caroliniensis.*
Diospyros virginicus.
Olea americana.
Ilex vomitoria.
Pines.
Oaks.
Cypress (*Cupressus disticha*).
Tillandsia usneoides.
Fan palm (*Chamærops palmetto*).
Salix nigra.
Populus deltoides.
Laurus sassafras.
Myrica caroliniensis.
Miegia macrosperma.
Ludolphia mississipensis.
Rubus.
Vitis riparia.
Ampelopsis bipinnata.
Juglans pacan.
„ *rubra.*
Acer negundo.
Sarracenia.
Venus's fly-trap.
Rhododendrons.
Azaleas.

Cultivated.

Cotton.
Maize.
Rice.
Wheat.
Sugar-cane.
Indigo.
Tobacco.

WARMER TEMPERATE ZONE OF NORTH AMERICA.

(From 34° to 45°.)

Wild.

Solidago.
Asters.
Oaks.
Pines.
Vaccinæ.
Liriodendron tulipifera.
Mimosa.

Smilax china.
„ *hastata.*
Smalteri.
Miegia macrosperma.
Ludolphia mississipensis.
Bignonias.
Chestnut.
Ashes.
Beech.
Elms.
Hazel.
Plantanus occidentalis.
Roses.
Walnut.
Wellingtonia gigantea.
Sugar maple.

Cultivated.
Maize.
Wheat.
Barley.
Oats.
Rye.
Tobacco.
Flax.
Hemp.
Vine.
Hops.
Apple.
Sugar-cane.
Beans.
Water-melons.
Peach.
Buckwheat.
Potato.
Cotton.

Colder Temperate Zone of North America.

(From 45° to 58°.)

Wild.
Firs.
Alder.
Sorbus.
Hawthorn.
Rubus odoratus.
Cornus suecica.
Gooseberry.
Azalea.
Calla.
Linnæa borealis.
Lathræa stelleri.
Panax horridum.
Arenaria peploides.
Glaux maritima.
Carex.
Juncus.
Veronica serpyllifolia.
„ *anagallis.*
Potentilla anserina.
„ *ruthenica.*
Sisyrinchium.
Plantago.
Triglochin.
Dodecatheon.
Pedicularis.
Elymus.
Bartsia.
Campanula.
Angelica.
Heracleum.
Fritillaria.
Pisum maritimum.
Cochlearia danica.
Ranunculus acris.
Galium boreale.

Cultivated.
Maize.
Tobacco.
Wheat.
Barley.
Oat.
Peas.
Rye.
Hemp.
Potato.

Sub-arctic and Polar Zones.

(From 58° to the North Pole.)

Wild.

Rhododendron lapponicum.
Andromeda tetragona.
„ *polifolia.*
„ *caliculata.*
Vaccinium vitis idœa.
Oxycoccus palustris.
Azalea procumbens.
Birch (*Betula glandulosa*).
Alder (*Alnus glutinosa*).
Saxifraga aizoides.
„ *oppositifolia.*
„ *cernua.*
„ *groenlandica.*
Polygonum viviparum.
Arnica montana.
Dryas integrifolia.
Holcus alpinus.
Pedicularis lapponica.
„ *sudetica.*
„ *hirsuta.*
Plantago lanceolata.
Cerastium viscosum.
Oxyria reniformis.
Triglochin maritimum.
Tofieldia borealis.
Epilobium palustre.
„ *latifolium.*
„ *angustifolium.*

West India Islands.

Wild.

Mahogany (*Swietenia Mahogani*).
Pimento (*Myrtus Pimenta*).
Avocada pear (*Laurus Persea*).
Papaw tree (*Carica Papaya*).
Anacardium occidentale.
Passion-flowers.
Pine-apple.
Mammee tree (*Mammea americana*).
Mango.
Teak.
Mimosa.
Trumpet tree.
Banyan fig.
Tree ferns.
Heliconias.
Cocoa-nut.
Date.
Cabbage palm.
Nutmeg.
Cassia.
Cannon-ball tree.
Screw pine (*Pandanus odoratissima*).
Bread fruit tree
Quercus suber.
„ *robur.*
Piper nigrum.
Cactus.
Euphorbia.
Bamboo.
Ferns.
Orchids.

Cultivated.

Banana.
Pine-apple.
Sugar-cane.
Coffee.
Cotton.
Maranta arundinacea.
Pimento.
Cocoa.
Indigo.
Ginger.

SOUTH AMERICA.

Brazil.

Wild.
- Cassias.
- Cecropias.
- Myrtles.
- Cacti.
- Bignonias.
- Paullinias.
- Passion-flowers.
- Coronilla.
- Macauba-palm.
- Securidaceæ.
- Mikanias.
- Celtis.
- Rhexias.
- Melastomas.
- Bauhinias.
- Mimosa.
- Nightshades.
- Sebestanas.
- Eupatoria.
- Crotons.
- Cerei.
- Silk cotton tree.
- Brazil wood tree.
- Lecythis.
- Cabbage palm.
- Begonias.
- Costus.
- Heliconias.
- Brazilian pine (*Araucaria imbricata*).
- Mangrove tree (*Rhizophora Mangle*).
- Laurel.
- Geoffræa.
- Barbadoes cedar.
- Eugenia.
- Theophrasta.
- Conchocarpus.
- Dwarf-palms.
- Bindweed.
- Pothos.
- Arum.
- Orchids.
- Bromelias.
- Ferns.
- Brazil nut (Bertholletia excelsa).
- Cocoa or chocolate tree.

Cultivated.
- Orange.
- Coffee.
- Sugar-cane.
- Maize.
- Beans.
- Yam.
- Batatas.
- Cotton.
- Tobacco.
- Mandiocca root (*Jatropha Manihot*).

Colombia.

Wild.
- Banana.
- Heliconia.
- Alpinia.
- Theophrasta.
- Plumieria.
- Cæsalpinia.
- *Cecropia peltata.*
- Balsam tree of Tolu.
- Quinine tree.

Mangrove tree.
Ceroxylon andicola.
Tree ferns.
Cinchona lancifolia.
„ *cordifolia.*
„ *longifolia.*
„ *oblongifolia.*
Fuchsias.
Passion flowers.
Sisyrinchium.
Melastomæ.
Bocconia frutescens.
Arums.
Quercus granatensis.
Clusias.
Calceolarias.
Lobelias.
Walnut tree.
Ebony tree.
Sassafras.
Cocoa tree.
Vanilla.
Tamarind.
Medlar.
Manchineel tree.
Cassia.
Sapotas.
Guava.
Arracacha (*Arracacia esculenta*).
Cow tree (*Galactodendrum utile*).

Cultivated.
Cocoa.
Tobacco.
Coffee.
Cotton.
Sugar.
Banana.
Batatas.

PERU.

Wild.
Cacti.
Tillandsia.
Melocacti.
Cerei.
Willow.
Manglillo.
Huarango.
Cordia.
Buddlea.
Heliotropium.
Lantana.
Lycium.
Schinus Molle.
Potato.
Cocoa-nut palm.
Orchid.

Cultivated.
Yellow potato of Peru (*Papas amarillus*).
Maize.
Vine.
Sugar-cane.
Barley.
Wheat.
Rye.

LA PLATA.

Wild.
Clover.
Ombu (*Phytolacca dioica*).
Thistles.
Cardoon (*Cynara Cardunculus*).
Ferns.
Curily pine (*Araucaria brasiliensis*).

Ybaro (*Sapindus saponaria*).
Cacti.
Paraguay tea (*Ilex paraguensis*).
Leguminous plants.

A great portion of the country is occupied by pampas, extensive treeless plains, covered with grass.

Cultivated.
- Vine.
- Tobacco.
- Sugar.
- Mandiocca.
- Maize.
- Batatas.
- Wheat.
- Barley.
- Paraguay tea.
- Peach tree.

Chili.

Wild.
- Roble (*Fagus obliqua*).
- Lingui (*Laurus lingui*).
- Queule (*Gonmertiganitida*).
- Laurel (*Laurelia aromatica*).
- Canelo (*Drymis chilensis*).
- Avellano (*Quadria heterophylla*).
- Chili pine.
- Boldo (*Boldoa fragrans*).
- Peumo (*Peumos rubra*).
- Fuchsia.
- Mimosa.
- Cavenia.
- Palm of Chili (*Molinæ micrococcus*).
- Saccharinæ.
- *Acacia Caven.*
- *Ephedra chilensis.*
- *Myrtus communis.*
- Calceolarias.
- Baccharis.
- Eupatorium.
- Proustia.
- Mutisiæ.
- *Lobelia tupa.*
- Psoralea.
- Cestrum.
- Dodder.
- Salpiglossis.
- Malesherbia.
- *Loranthus aphyllus.*

Cultivated.
- Wheat.
- Potato.
- Pumpkin.
- Lettuce.
- Cabbage.
- Vine.

Terra del Fuego and Straits of Magellan.

Wild.
- Evergreen beech (*Fagus betuloides*).
- Birch (*Betula antarctica*).
- Fagus antarctica.
- Cranberry.
- Drymis Winteri.
- Wild celery.
- Scurvy grass.
- Veronica.
- Fuchsia.

POLYNESIA.

Wild.

Pandanus odoratissima.
Dracæna terminalis.
Morus papyrifera.
Erythrina corallodendron.
Santalum Freycinetianum.
Bamboo.
Acacia heterophylla.
Pine apple.
Casuarinæ.
Aleurites triloba.
Ipomea.
Ananas.
Asplenium nidus.
Pteris.
Piperaceæ.
Lobelias.
Urticeæ.
Charpentiera obovata.
Metrosideros polymorphus.
Jambosa Malaccensis.

Both wild and cultivated.

Bread fruit (*Artocarpus incisa*).
Cocoa-nut (*Cocos nucifera*).
Yam (*Dioscorea alata*).
Taro (*Arum esculentum*).
Plaintain(*Musa sapientum*).
Banana(*Musa paradisiaca*).
Mape or rata (*Inocarpus edulis*).
Brazilian plum (*Spondias dulcis*).
Ahio (*Eugenia malaccenus*).
Pteris esculenta.
Polypodium Medulla.
„ *dichotomum.*

AUSTRALIA AND NEW ZEALAND.

Wild.

Proteas.
Acacias.
Eucalyptus.
Epacris.
Myrtles.
Leguminous plants.
Composite plants.
Melaleuca.
Gnaphalium.
Casuarina.
Mimosa.
Araucaria excelsa.
Orchids.
Callitris.
Xanthorrhea.
Doryanthes excelsa.
Cephalotus follicularis.
Castonospermum australe.
Claudea elegans.
Phormium tenax.
Banksia.
Dryandra.
Bœckia.
Leptospermum.
Cupressus callitris.
Olax.
Exocarpus.
Angophora.
Metrosideros.
Zamia spiralis.
Drosera pedata.
Hœmodorum teretifolium.
Ferns.
Marsdenia suaveolens.
Stackhousia monogyna.
Samolus littoralis.
Hibbertia volubilis.
„ *diffusa.*
Juncus vaginatus.

Lycopodia.
Loranthus.
Kanguru-grass (*Anthristiria australis.*)
Mesembryanthemum æquilaterale.
Polygonum junceum.
Viscum.
Lobelia.
Dichondra repens.
Epilobium.
Alisma.
Triglochin.
Actinocarpus.
Najas.
Lemna.
Cyperus.
Scirpus.
Schœnus.
Carex.
Myriophyllum.
Mentha.
Dracœna australis.
Areca olerace

THE END.

LONDON:
SAVILL AND EDWARDS, PRINTERS, CHANDOS STREET,
COVENT GARDEN.